tledge Studies in Technology, Work and Organizations

ED BY DAVID PREECE, *UNIVERSITY OF TEESIDE, UK*

formation Society and the
orkplace
dited by Tuula Heiskanen
d Jeff Hearn

chnological Change and
ganizational Action
lited by David Preece
d Juha Laurila

naging Complex Projects
tworks, Knowledge and
ovation
l Alderman, Chris Ivory,
McLouglin, Alfred Thwaites
Roger Vaughan

rmation and Communication
nnologies in Rural Society
ng Rural in a Digital Age
e Rusten and Sarah Skerratt

ware and Organizations
Biography of the Enterprise-
System or how SAP
quered the World
Pollock and Robin Williams

nological Communities and
orks
ers and Drivers for Innovation
ris Assimakopoulos

h Technology Development
Jse
Practice-Bound Imagination
olving Impacts
sa Hyysalo

8 Nanotechnology and Sustainable Development
Claire Auplat

9 Challenging the Innovation Paradigm
Edited by Karl-Erik Sveiby, Pernilla Gripenberg and Beata Segercrantz

Challenging the Innovation Pa

Rou

Edit

1 In
W
E
ar

2 Te
Or
Ed
an

3 Ma
Ne
Inn
Ne
Ian
and

4 Info
Tec
Bei
Gre

5 Soft
The
wide
Con
Neil

6 Tech
Netw
Trigg
Dimi

7 Healt
and U
From
to Ev
Samp

Challenging the Innovation Paradigm

Edited by Karl-Erik Sveiby, Pernilla Gripenberg and Beata Segercrantz

Routledge
Taylor & Francis Group
NEW YORK LONDON

First published 2012
by Routledge
711 Third Avenue, New York, NY 10017

Simultaneously published in the UK
by Routledge
2 Park Square, Milton Park, Abingdon, Oxon OX14 4RN

Routledge is an imprint of the Taylor & Francis Group, an informa business

Library of Congress Cataloging-in-Publication Data

Challenging the innovation paradigm / edited by Karl-Erik Sveiby, Pernilla Gripenberg and Beata Segercrantz.
p. cm. — (Routledge studies in technology, work and organizations ; 9)
Includes bibliographical references and index.
1. Technological innovations—Economic aspects. 2. Diffusion of innovations. 3. Research, Industrial. I. Sveiby, Karl Erik. II. Gripenberg, Pernilla. III. Segercrantz, Beata.
HC79.T4C423 2012
338'.064—dc23
2011043890

ISBN13: 978-0-415-52275-5 (hbk)
ISBN13: 978-0-203-12097-2 (ebk)

Typeset in Sabon
by IBT Global.

Contents

Figures

Tables

Series Editor Introduction

'Challenging the Innovation Paradigm'—Routledge Studies in Technology, Work and Organizations

When Pernilla Gripenberg asked me if I would like to write an Introduction to the present book, I was only too pleased to say 'Yes,' in part because it occurred to me that this would provide me with an opportunity to reflect on the series, and in part because I believe that this is a most significant and potentially ground-breaking addition to the series.

Regarding the series: the word 'series' needs to be interpreted in the plural here, for the present one is actually the third that I have edited for Routledge in the technology, organizations and work domain. The first series, entitled *Routledge Studies in the Management of Technology*, was jointly edited by myself and John Bessant and was launched in 1995 with the publication of my sole-authored book *Organizations and Technical Change: Strategy, Objectives and Involvement*; four further books followed. In 1997, the title of the series was changed to *The Management of Technology and Innovation*, and I became the sole editor. Seven books were published between then and 2000, when we launched the current series with another title change to *Routledge Studies in Technology, Work and Organizations*; this is the seventh book to be published in the series, and another book is contracted and others are in the 'pipeline.' Including the contracted book, this gives a total of twenty books published in the series (plural) over a period of 16 to 17 years. We have clearly hit on an important area of interest for the academic-researcher, practitioner, and policy-making fraternities!

The series throughout has focused on publishing books that critically explore some particular area within the technology, organizations, and work domain, drawing on and developing cognizant theorizations and providing and analyzing new empirical material that is both informed by and informs relevant theory. We have also always encouraged authors researching in the area based both in and outside of the UK to submit proposals for books, and I am pleased to say that series authors/editors include people based in England, Scotland, Finland, Australia, Sweden, Norway, and France. The changes to the series titles are indicative of how the focus has changed from the early one on (but by no means exclusively) technology management and change and intraorganizational material and theorizing, to a wider focus on technology and work (as well as organizations), in other words, technology in interorganizational and nonorganizational settings (e.g., in the home). Other key changes have been to deemphasize (but by no means exclude!) technology management and to welcome contributions that examine technology development, use, deployment, and consequences/implications for a range of other stakeholders, such as nonmanagerial employees, users, customers, consultants, and suppliers.

Turning to the present book, this is timely, significant, and in keeping with the series aims, that is, we have here both a good deal of new empirical material, a critical summary of and reflection on the relevant extant literature, and a significant development of that literature, in many respects emerging from a reflective engagement with the new material gathered by the authors and collectively presented here for the first time. We are long overdue a book that we as researchers, teachers, or policy makers/administrators can draw on and recommend to others (not least students) as an antidote to what the editors/authors refer to as a 'pro-innovation bias' in the great majority of the previous and present work in the area. Too many matters, aspects, and consequences (especially the unintended and undesirable ones) of innovation have been taken for granted and left unexamined. Highly deleterious effects have occurred in many cases, as illustrated in the book, that might have been avoided if there had been a more sanguine treatment of innovation ideas and proposals at an early stage.

Technology management writers—including myself—have long advocated that technological/organizational changes are evaluated and reflected on through some form of organizational mechanism to learn from what has happened during and as an outcome of the change process, how it has happened, why it has happened, and so on, with a view to 'doing things differently next time,' that is, through organizational learning. Unfortunately, we also know that, for a variety of reasons, this does not occur, or at best in only a cursory way. This can be applied equally well in the innovation field, but also in the early days, before the innovation has materialized; however, for a reasonable chance of this to happen, the actors involved need to take a much more agnostic and critical stance toward any such ideas or proposals than they typically do now. The present book examines and explains why this is so, and, if its message is disseminated as we hope, it will contribute to that process.

Professor David Preece
Head of Centre for Leadership & Organizational Change
The Business School
Teesside University
September 2011

Preface

This book is the result of a project that emerged as a response to concerns about unintended and undesirable consequences of innovation among a group of researchers at the Department of Management and Organisation at the Hanken School of Economics. Inspired by a lively dinner conversation at the departmental Christmas party in 2008, we decided to pursue this question more in-depth. What we found was that although innovation is one of the most commonly mentioned concepts in social science, unintended undesirable consequences of innovation are rarely studied. This was the starting point of a project to address this shortcoming and the implications from it.

The first research task of the project was to study research on unintended undesirable consequences of innovation. Our aim was to see how much of innovation research today captures or explores undesirable effects of innovations, what subjects are being covered in this research, what kind of consequences are found, who are affected by these consequences, and what methodologies are being used to study consequences of innovation. A further aim was to get a tentative answer to the question 'Why are so few innovation researchers interested in the consequences of innovation?' We conducted two empirical studies: a review of articles in peer-reviewed journals with innovation, new product development (NPD), or similar notions in the title and a survey of corresponding authors to articles in the innovation research field. The results were first presented in a paper at the International Society for Professional Innovation Management (ISPIM) conference in Vienna in June 2009, and also at a Hanken research seminar in September 2009.

The paper raised interest and discussion although some participants had difficulty seeing the underlying aim of the research. It seemed as if we were not able to express our concerns clearly enough to colleagues in the mainstream innovation field. We then shortlisted a group of international scholars that had shown an interest in the issues raised in the paper or who had themselves taken a critical approach to innovation in recent papers. These scholars were invited to a workshop held at the Hanken School of Economics in May 2010 with the title "Beyond the Pro-Innovation Bias.' This book is an offspring of that workshop.

The aim of the workshop was to bring together a group of fifteen to thirty innovation researchers to explore alternatives to mainstream innovation research by addressing how unintended and undesirable consequences of innovation could be brought into the research agenda of the innovation research field. We wanted to go beyond the 'pro-innovation bias.' Over twenty researchers from a dozen countries took part in the workshop. Another important aim was to start the production of a publication. Inspired by two keynote speakers, Nancy Harding and Jan Fagerberg, the participants worked in small groups throughout the workshop, brainstorming, planning, and even writing what was to become the current book.

Bringing the project this far means that we are indebted to a number of people and organizations. First, we would like to thank Andreas Erikson and Alexander Aminoff, who, under our firm but hopefully constructive supervision, conducted the first two studies as part of their master theses. Second, and perhaps most important, we would like to thank all the participants in the workshop. Rarely can you find such energy, creativity, and commitment as we did at the workshop in Helsinki in May 2010. In relation to the workshop, we also wish to thank the keynoters Jan Fagerberg and Nancy Harding for inspiration. We also want to express our gratitude to David Preece, the series editor at Routledge, for his very constructive support right from the beginning of the book writing process. Special thanks also go to the section editors of this book, who helped us keep track of how the work was proceeding and deadlines met. Last but not least, we want to thank all the contributors to the book, some of which have been with us since the workshop, others who have joined in later: Thank you for your commitment to produce the insightful, inspiring, and exciting chapters!

Helsinki, February 8, 2012
Karl-Erik Sveiby, Pernilla Gripenberg, and Beata Segercrantz

1 Challenging the Innovation Paradigm

The Prevailing Pro-Innovation Bias

Pernilla Gripenberg, Karl-Erik Sveiby, and Beata Segercrantz

Currently innovation is not only seen as a key driver of organizational success but as vital for organizational survival. At a societal level, innovation is perceived as one of the key drivers of economic growth—growth that leads to national wealth and prosperity—and therefore innovation is currently often regarded as the ultimate solution to present welfare related problems in the West (see for example, Organisation for Economic Co-operation and Development (OECD), 2010). As a result, there is a constant strive to innovate, particularly in Western societies. Perhaps surprisingly, whereas innovation is one of the most commonly mentioned concepts in social science, unintended undesirable consequences of innovation are rarely studied. On the other hand, it is perhaps not surprising as it is typically taken for granted that 'innovation is always good' and this assumption remains widely accepted and unquestioned. In his review of innovation literature 25 years ago, Rogers (1983) noticed that only 0.2% of innovation research articles addressed *consequences* of innovation. To follow up this study, we conducted a study including a literature review of all articles in the EBSCO database (see chapter 4) with innovation in the title and that study undesirable consequences. Although 'innovation' hits reach hundreds of thousands, refining searches into 'negative or undesirable consequences' decreases the hits radically. The study found only 26 articles on unintended and undesirable consequences of innovation; 1 per 1,000, a proportion that has not changed since the 1960s.

And still today, the study of consequences thus seems to be marginalized across all main areas of the rapidly emerging scientific field of innovation. Fagerberg and Verspagen (2009) summarize the three main areas of this field of research: the role of innovation on long-term economic growth, the contexts of innovation, and the national systems of innovation. Typically studies within these different areas of innovation research treat innovation as the dependent variable, whereas studies investigating innovation as the independent variable are uncommon (Anderson, De Dreu, & Nijstad, 2004). Accordingly, the study of consequences and *unintended* and *undesirable* consequences of innovation in particular are rare.

Innovation research seems to be built on a dominant assumption that 'innovation is good' regardless of the consequences. This is also known as the 'pro-innovation bias' (Abrahamson, 1991; Kimberly, 1981; Rogers, 1983). This bias limits the ability of decision makers and change agents to anticipate unintended and undesirable consequences. A way to understand how the bias is continuously reinforced is by considering how 'innovation' is socially constructed on various levels of society: How is innovation talked and written about by various authorities, such as, policy makers, funding bodies, and researchers? What assumptions constitute discourses of innovation? What are the consequences of these assumptions for how innovation research is conducted and focused? It seems important to go beyond the pro-innovation bias and current dominant discourses and ask why and how certain understandings of innovation are constructed and maintained, what the implications are, and how excluded or silenced voices can be promoted in the research field of innovation. These are questions that this book addresses.

The pro-innovation bias has already been recognized and discussed since the 1970s (Downs & Mohr, 1976; Kimberly, 1981; Rogers, 1983; Rogers & Shoemaker, 1971; Van de Ven, 1986; Zaltman, Duncan, & Holbeck, 1973). Both Kimberly (1981) and Rogers (1983) agree that one of the reasons for this 'one-sidedness' of innovation research is historical: Innovations have had such positive outcomes on economic growth in the U.S. after World War II that other possible outcomes have remained marginal and thus received little attention. Moreover, this bias is still with us and reflects the current state of innovation research. Research, which builds on the assumption that innovation primarily has positive results, seeks to define situations where innovation occurs and where and how innovations spread more easily and quickly. A further focus is put on what characteristics distinguish innovative individuals, groups, or organizations from less innovative ones. An implicit assumption that research should promote innovation also seems to prevail.

New ideas that are not perceived as useful are often seen as mistakes and thus disregarded rather than seen as failed innovations worthy of further consideration in terms of, for example, why they failed (Van de Ven, 1986). The same disinterest prevails in 'anti-diffusion' efforts, for example, to prevent the diffusion of 'bad' innovations such as drugs or cigarettes (Rogers, 1983) and 'exnovation': the rejection or discontinuance of an innovation (Kimberly, 1981). This disinterest by innovation researchers in what happens after an innovation is diffused and implemented (especially if it fails) can be taken as a general indication of the prevailing one-sidedness of the innovation literature (Downs & Mohr, 1976; Kimberly, 1981; Van de Ven, Angle, & Poole, 2000, p. 13).

The book thus attempts to bridge the prevailing separation of discourses on desirable and undesirable consequences of innovation. More specifically, undesirable consequences of innovation can potentially follow from all types of innovation. These consequences are to some (unknown) degree studied in disciplines such as biology, medicine, environmental studies, and sustainable development, and theories are constructed within perspectives drawing on sociology, critical management studies (CMS), science and technology studies (STS), and so on. Innovation research, however, seems to have implicitly isolated itself from considering them (with notable exceptions, such as the special issue in *Technology Analysis & Strategic Management*, 2000, vol. 12, no. 3). The separation of discourses in this manner can have severe implications, as is further explicated in chapter 4.

In light of the issues previously raised, the field of innovation research is in need of both theoretical and methodological development. Undesirable consequences of innovation are notoriously complex, multifaceted, and multilevel, and they transcend artificially created scientific boundaries. They require multidisciplinary research and cross-boundary dialogue. They are difficult, but not impossible, to study.

AIM OF THE BOOK: CHALLENGING CONTEMPORARY INNOVATION RESEARCH

The aim of this book is to challenge contemporary innovation research by problematizing its underlying assumptions and promoting a more nuanced way of considering the consequences of innovation. The questions we pose are: How is 'innovation' and its underlying assumptions constructed; what are the consequences of these; and what does this mean for society? As the book is a first attempt at this kind of research, we can only scratch the surface of this topic, and have thus chosen three focus themes to study more in-depth. The first is to explore a variety of national and supranational discourses on innovation; the second is to illustrate the systemic nature of innovation and its possible acceleration through cases; and the third is to explore unintended consequences of innovation in particular contexts.

The book thus contributes to our understanding of how the dominant discourse of innovation is constructed and reconstructed at different levels of society—a necessary prerequisite for deconstructing these debates. As the book focuses on the consequences of innovation, it provides illustrations of how both the discourse of innovation and the consequences of innovation permeate all levels of society.

The book both reflects on and attempts to answer the following questions:

1. How is innovation discursively constructed? What consequences does this have on, for example, inclusion and exclusion, ethical considerations, and the overall ability of change agents (policy makers, innovators, funders, researchers, entrepreneurs, etc.) to consider unintended and unanticipated consequences of innovation? In particular the book looks into the following areas:

- academic and policy discourses of innovation (Chapter 2)
- the 'origins' and the development of innovation as a contested category over time (Chapter 3)
- the current state of research on unintended and unanticipated consequences of innovation (Chapter 4)

2. Given the systemic nature of innovation, is the speed of innovation accelerating in contemporary society, and what are the consequences? What kinds of counterforces are there to the innovation frenzy? The following questions are posed:

- Is the speed of innovation accelerating, and what are the potential consequences of acceleration? (Chapter 5)
- How does radical and accelerating innovation affect society in terms of risks, for example, in the context of the financial sector? (Chapter 6)
- There are weak signals of decelerating innovation in a few areas. What are these, and what can we learn from them? (Chapter 7)

3. What are the unintended consequences of innovation on society? The following questions are explored:

- How are negative consequences included in organizational models of innovation? (Chapter 8)
- How do rapid and increasingly formalized processes of innovation affect ICT experts in the ever-restructuring software industry? (Chapter 9)
- How does negative consequences affect human well-being, for example, in the organizational and work context, and what indirect consequences may follow from these? (Chapter 10)
- How do negative consequences affect the environment, for example, in the case of CO_2 emissions? (Chapter 11)

DEFINING INNOVATION AND CONSEQUENCES OF INNOVATION

On various occasions in the course of this project we have been asked what exactly we mean by 'innovation.' It is a valid, but difficult question to answer, as we are interested in how 'innovation' is constructed (that is, defined and not defined by others) and consequences of innovations that are not commonly associated or even regarded as innovation (that is, undesirable consequences). This section briefly discusses the term 'innovation' and what it may mean in mainstream innovation research. The section continues to briefly outline a summarizing framework of the theoretical foundations on consequences of innovation that inspired us to undertake the project that this book is a result of. It may not be the theoretical foundation for all the individual chapters in this volume, but it is a part of the common ground from where the project started in early 2009.

Rogers (1983, p. 11) defines innovation as 'an idea, practice, or object that is perceived as new by an individual or other unit of adoption.' Similarly, Van de Ven (1986, p. 591) defines innovation as: ' . . . a new idea, which may be a recombination of old ideas, a scheme that challenges the present order, a formula, or a unique approach which is perceived as new by the individuals involved.' An innovation can thus be new to one person and old news for someone else, as it takes time for innovations to be communicated and spread through a social system (Rogers, 1983). In accordance, Kline and Rosenberg (cited in Fagerberg, 2006, p. 5) emphasize that innovations should not be treated as well-defined, homogenous things 'that could be identified as entering the economy at a precise date.' Further, innovations go through drastic changes during their life span that may totally transform their significance in society. Thus the term 'innovation' not only lacks specific definition, but is also fluid: innovations cannot be specified in time, and they may be in constant transformation. They are also often the result of a long process involving many interrelated innovations (Fagerberg, 2006).

An example can be found in Ahmed and Shepherd's recent textbook *Innovation Management* (2010, p. 5), where they discern a number of characteristics in the multiple meanings given to innovation. Innovation can be understood as: creation, diffusion and learning, an event, a trajectory, change (incremental or radical), (firm-level) process, and context (region, nation, etc.) level process. Further, innovation can be seen as a value-adding process as well as an outcome. Ahmed and Shepherd also recognize the problem of newness in that, as an example in product innovation, an innovation can be new to the world, new to a firm, a new product line extension, a product improvement, or a product repositioning. A distinction between invention and innovation is also sometimes made so that invention stands for the generation of a new idea, whereas innovation stands for the fruitful commercialization/exploitation/market capitalization of that idea (for example, Ahmed & Shepherd, 2010).

The OECD provides its own definition in the Oslo Manual for collecting and interpreting innovation data on a national basis. Innovation is here defined as the implementation of a new or significantly improved product (good or service), or process, a new marketing method, or a new organizational method in business practices, workplace organization, or external relations.

> 'By definition, all innovation must contain a degree of novelty. The Oslo Manual distinguishes three types of novelty: an innovation can be new to the firm, new to the market or new to the world. The first concept covers the diffusion of an existing innovation to a firm—the innovation may have already been implemented by other firms, but it is new to the firm. Innovations are new to the market when the firm is the

first to introduce the innovation on its market. An innovation is new to the world when the firm is the first to introduce the innovation for all markets and industries' (OECD and Eurostat, 2005, in Organisation for Economic Co-Operation and Development, 2010).

In the beginning of this project, we were interested in innovation research as a research field. We therefore needed quite a different approach than the previous definitions to establish what counts as innovation research. Our definition was then simply to regard research articles that have the word 'innovation' in the title, abstract, or key words as 'innovation research.' Subsequently, authors that have written such an article count as an 'innovation researcher.' In other words, we left it to the authors in the field to do the defining. This distinction of course leaves out research that, for example, is considering severe, long-term consequences of innovation, such as climate change or nuclear waste disposal, as this research rarely links back to problematize the original innovation per se.

This is precisely one of the key points we want to highlight in this book: that discourses on innovation and on consequences have been dangerously separated. That is, those who innovate and support innovation do not necessarily see the need to take responsibility for or even consider the consequences. Four theories that consider the consequences and particularly the unintended consequences of innovation inspired us in the initiation of the research project that produced this book: economic theory (Schumpeter, 1939; Smith, 1789), communication theory (*Diffusion of Innovation*) (Rogers, 1983), Merton's (1936) sociological theory, and stakeholder theory (Freeman 1984).

The theoretical framework of the consequences of innovation is elaborated in chapter 4. A summary is presented in Figure 1.1. For each

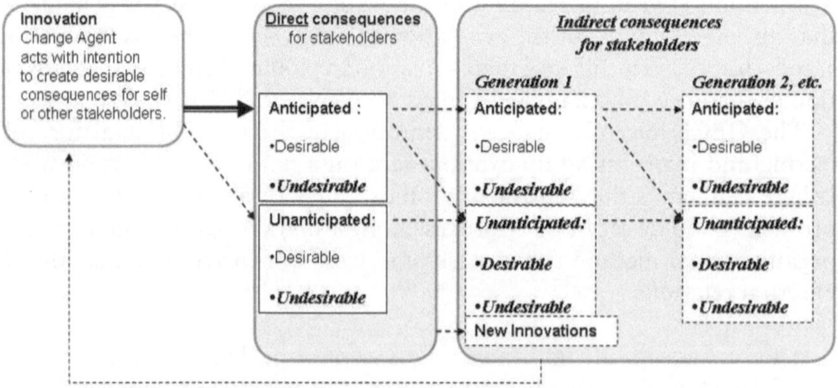

Figure 1.1 Consequences of innovation.
Source: Building on Freeman (1984); Merton (1936); Rogers (1983); and Schumpeter (1939).

anticipated and desirable direct consequence of change or innovation, there are numerous other consequences: anticipated and unanticipated, desirable and undesirable, and indirect in several generations. Often one innovation cause further innovations (Schumpeter, 1939), forming a feedback-loop in the model. The model thus indicates that long-term, there are more unanticipated than anticipated consequences and that innovation is *systemic*. The claim this book makes is that contemporary innovation research and the innovation discourse in general is overly focused on: first, the earliest stage, what leads to an innovation; and second, on only one type of consequences: those that are *direct*, *anticipated*, and *desirable*, whereas other types of consequences are largely subdued or neglected entirely. In Figure 1.1, the neglected areas are highlighted in bold italics. Given the systemic nature of innovation, this is a rather worrying situation, and it is this gap and the reasons for it that this book is addressing in more detail.

STRUCTURE AND CONTENT OF THE BOOK

In the first section of the book, our effort is to problematize current conceptualizations of innovation by unpacking the historical origin and development of the category of innovation toward its present state as a nonproblematic, de-contested ideology. In the first chapter of the section, Fougère and Harding track the development of three texts that can be considered seminal in the articulation of 'innovation.' They explore how innovation in the late 20th century has come to be understood as something desirable and measurable to the extent that it is impossible to think about innovation outside these terms. The texts are: Rogers's 1962 book *Diffusion of Innovations* and the four consecutive editions (1971; 1983; 1995; 2003); the 1967 Charpie Report; and the following Oslo Manuals (1992; 1997; 2005). The chapter shows how Rogers' texts, by relating failures of innovation only to the 'non-West' and cultures that have kept a very traditional livelihood (aborigines, Amish), establish an absolute opposition between the non-innovative communities and 'us' who try to help them. Hence, the texts have come to reinforce the view that to innovate is to be modern and advanced, and to be modern is to be Western: innovation has become an aspect of Western identity.

Further, the chapter shows how the original Charpie Report and the Oslo Manuals attempt to bring innovation under the control of managers by developing measures and ways to count innovation. As time has passed, this has become rewritten to pass innovation off as founded in science and rationality, something that was not in the original Charpie Report. Taken together then, this has lead to the current understanding that being non-innovative is to be 'backward,' and because failed innovations cannot be counted, they disappear from the data, that is, they are not innovations. Failing innovations and undesirable consequences are not part of Western identity. Failure is what happens when 'others' (non-West, premodern cultures), 'unable to handle' innovation, try to innovate.

However, as Godin shows in the following chapter, that has not always been the case. On the contrary, until the 20th century, for at least the past 2,500 years, to innovate was pejorative; it was forbidden and represented as something negative. Historically, innovation meant introducing change into the established order, and, as such, it was forbidden. Gradually, the meaning of innovation changed to mean novelty of any kind and was also used to talk about a new era, but it still had nothing to do with technology or commercialization.

It was not until the turn to the 19th century in conjunction with the shift in conception of time and reorientation toward the future that innovation was being discussed for the first time in more positive terms, as a break with the past. In science, innovation first meant novelty in methods, not technological innovation, and then gradually it came to mean revolutionary novelty: something absolutely new that has never existed before. The chapter further shows how it was only in the 20th century, when innovation was no longer seen merely as change, that concepts such as 'creativity,' 'utility and usefulness,' 'commercialization,' and 'process' were being linked to the understanding of innovation. This was largely due to theorists in sociology, economics, anthropology, and history. Today, '[t]o governments and organizations, innovation is technological innovation and is a tool of policy, a tool in the service of industry. The rationale is essentially economic: competitiveness between industries and between countries.' But, until the 19th century, 'YOU are the innovator' was certainly not something one wanted to be accused of.

In the last chapter of this section Sveiby, Gripenberg, and Segercrantz show first how consequences of innovation have been theorized in the 20th century, in particular in some influential writings in economic theory (Schumpeter), communication theory (Rogers), a sociological theory (Merton), and stakeholder theory (Freeman). The chapter then goes on to review current innovation literature only to find that research on unintended and undesirable consequences of innovation is almost nonexistent, as it was also in the 1960s when Rogers did his first review on the same issue.

In the second section of the book, the effort is to understand the current innovation imperative and the systemic nature of innovation in more detail. We do this by exploring the accelerating speed by which innovation occurs; its reasons; its consequences; and its counterforces.

Hasu, Leitner, Solitander, and Varblane explore the dynamics and impact of what they call the speed imperative, which is embedded in the current, dominant understanding of innovation as a race or competition between firms, regions, nations, cities, the West, and other regions. Hasu and colleagues argue that in this race there is little space for critical reflection, craftsmanship, deep involvement, and deep learning. They recognize a paradox in that: 'On the one hand the acceleration of innovation is socially desirable; it may help in developing regions to advance and become more prosperous, and address social challenges. On the other hand, the speed imperative is a mechanism of greed and unsustainability; speeding of the introduction of new product variations into markets enables global enterprises to make more profit.' In the chapter, Hasu and colleagues highlight a number of different areas where more time for reflection would reduce unintended and undesirable consequences of the speed imperative. They develop an integrated framework with more space for critical reflection, and they give examples of emerging practices to counter the speed imperative and thus slow down nonreflective, accelerating innovation.

In the following chapter, Sveiby demonstrates how the systemic nature of innovation, its rapid diffusion, in conjunction with the competence-destroying capabilities of innovation, may result in severe unintended and undesirable consequences with global effects. The chapter analyses a radical financial innovation, that of securitization, and follows its development through the financial instrument of Collateral Debt Obligations (CDO) over a period of four decades. The analysis is focused around critical events that diminish competences of central actors (issuers, bondholders, and agents in originating banks) in evaluating credit risk, price risk, liquidity risk, and model risk. The case shows how temporary incompetence lead to actors making prediction errors that then result in unintended consequences—in this case, global financial crises with severe effects not only for the financial industry, but for society at large. Sveiby uses a combination of massive quantitative and qualitative data to support his argument, and the case is both a thriller to read and an important wake-up call for the financial industry in particular, but also in general: due to the systemic nature of innovation and their consequences, faster innovation diffusion, particularly when innovations alter the context in which they are diffused, may not at all produce faster returns and benefits—on the contrary, it may be the faster route toward a major disaster.

Leitner then explores weak signals to identify a de-acceleration in the contemporary innovation pace. Based on a European foresight project on new ways to organize innovation in the future, he shows that although the intense innovation competition may be difficult to 'escape,' there exist 'weak signals' suggesting that some firms deliberately choose a no-innovation strategy. Such a strategy can take on different forms: it can be to avoid changing products to give them a traditional touch; it can be pseudo-innovation in the form of superficial adaptations of product design; it can be to apply for patents with the intention only to reduce competition; it can be to outsource innovation so that the firm can remain stable, whereas creative turbulence is transferred to sub-industries, -companies, or -groups of society.

Another trend is the aim toward sustainability, taking the form of eco-innovation, tending to the 'green market,' and taking a 'cradle-to-cradle' design approach where product components are circulated in biological and technical loops. Those and other similar strategies, which slow down the innovation pace, may result in better resource efficiency, more reflection, and environment and health improvements. Leitner also observes other weak signals that a number of new innovations nowadays come from companies that are located in countries such as China and India, that is, outside the currently leading innovators in the West. Interestingly, when compared to Godin (chapter 3), Leitner can already observe a trend toward a more positive attitude toward imitation and the adoption of imitation strategies. Will 'copycat' become the new identity for the West (cf. chapter 2)?

In the third section of the book, unintended consequences of innovation—both direct and indirect— are explored in relation to specific (empirical) contexts and specific innovations. First, Lindell reviews the innovation models that have been developed and disseminated in business literature over the past decades to explore to what extent they have included negative consequences for stakeholders other than company managers, owners, and customers; for instance employees and society in more general terms. He puts forward a number of suggestions that could remedy the lack of consideration of negative effects that the review found in the models, and an alternative innovation model is provided. The suggestions include legislation, top management actions, and an extension of innovation models to include the implementation phase, the diffusion phase, and the product growth and retirement phases, in addition to considering stakeholders much more broadly. This is much in line also with the eco-innovation and cradle-to-cradle models that Leitner suggests in chapter 7 and the variety of stakeholders affected by innovations as discussed by Sveiby and colleagues in chapter 4.

In the following chapter, Segercrantz explores how software experts are able to address unintended and undesirable consequences of the products they develop when the innovation process is formalized through organizational restructurings. In her study, she draws on discourse analysis to see how developer's subject positions change from being autonomous craftsmen to being compliant resources with a reduced sense of control over their work contributing to dis-identification. The formalization of software innovation processes seemingly have double negative consequences, first for the developers themselves who become somewhat removed and alienated from their own work, and second, this is bound to have consequences on the products they develop.

Next Cañibano, Basilio, and Sánchez explore the potential negative consequences of work intensification due to organizational innovations in human resource management (HRM) and in information and communication technology (ICT) for the well-being of employees. Work intensifies in the form of increased work time, greater effort, or both. Well-being is reduced due to "work overload, invasion of private life, high complexity and constant renewal that produce tension, stress, and even burnout.' The focus of their study is on flexible work arrangement and three new ICT tools that interfere with innovation, communication, and administrative tasks in a division of a major international consultancy firm. What they find is that organizational innovations produce work intensification because they distort work boundaries; induce multitasking; and accelerate the pace of work, and they require employees to constantly adapt, change, and learn. This is called involuntary work intensification, and it leads to greater stress, anxiety, and pressure, that is, reduced well-being. Cañibano and colleagues also find something else: When organizational innovations are introduced to empower and support employees voluntarily, work also intensifies, but this does not produce the same kind of stress and negative effects as with involuntary intensification. They also find that the adjustment period, the ability of the employee to disengage from work, and the proportion of time spent working alone compared to with colleagues affect the relationship between organizational innovation and decreased well-being. All in all, the chapter suggests that whether or not organizational innovations lead to negative effects for employees is largely a matter of managerial effort and concern.

Finally, in the third chapter of the section, Matsumoto and Kawajiri show in a rare study that it is possible to identify the *indirect* effects of ICT on CO_2 emissions—a task that seems impossible at first sight. With careful statistical analyses, focusing on different industrial sectors—textiles, electronics, and car manufacturing—combining data on energy consumption, production trends, and CO_2 emission trends over several decades in East Asia, they demonstrate, for example, that transferring electronic goods production from Japan to Korea and China actually increased the total CO_2 emissions. They demonstrate that whereas the direct effects of ICT innovations is to reduce CO_2 emissions, the innovations have also lead to a migration of production from industrialized to developing countries with much lower CO_2 efficiency. When this indirect negative effect is included, the total net effect is an increase in CO_2 emissions. The analysis is not only important in its findings, but it also provides an excellent example for the innovation research community of how indirect consequences of innovation can and should be studied.

The book concludes with a chapter where we draw together the findings presented in this book, bringing attention to the commercial waste, policy ineffectiveness, and human suffering caused by the way corporations have executed and policy makers have regulated innovation. Further, we consider the avenues for further research and the need for innovation research to shift the balance from its current positive bias to a balanced view on innovation and to contemplate the practical implications for various stakeholders of considering in more detail the unintended and undesirable consequences of innovation.

REFERENCES

Abrahamson, E. (1991). Managerial fads and fashions: The diffusion and rejection of innovations. *Academy of Management Review*, 16(3), 586–612.

Ahmed, P. K., & Shepherd, C. D. (2010). *Innovation management: Context, strategies, systems and processes.* Essex: Pearson.
Anderson, N., De Dreu, K. W., & Nijstad, B. A. (2004). The routinization of innovation research: A constructively critical review of the state-of-the-science. *Journal of Organizational Behavior*, 25, 147–173.
Downs, G. W., & Mohr, L. B. (1976). Conceptual issues in the study of innovations. *Administrative Science Quarterly,* 21, 700–714.
Fagerberg, J. (2006). Innovation: A guide to the literature. In J. Fagerberg, D. C. Mowery & R. R. Nelson (Eds.), *The Oxford handbook of innovation.* Oxford: Oxford University Press.
Fagerberg, J., & Verspagen, B. (2009). Innovation studies: The emerging structure of a new scientific field. *Research Policy,* 38, 218–233.
Freeman, R. E. (1984). *Strategic management: A stakeholder approach.* Boston: Pitman.
Kimberly, J. R. (1981). *Managerial innovation.* In I. P. C. Nystrom & W. H. Starbuck (Eds.), *Handbook of Organizational Design,* vol. 1, 84–104. New York: Oxford University Press.
Merton, R. K. (1936). The unanticipated consequences of purposive social action. *American Sociological Review,* 1, 894–904.
Organisation for Economic Co-Operation and Development. (2010). OECD Innovation Strategy: Innovation Strategy Key Findings. Retrieved May 10, 2010, from http://www.oecd.org/document/34/0,3343,en_41462537_41454856_41471586_1_1_1_1,00.html
Rogers, E. M. (1983). *Diffusion of innovations* (3rd ed.). New York: Free Press.
Rogers, E. M., & Shoemaker, F. F. (1971). *Communication of innovations: A cross-cultural approach.* New York: Free Press.
Schumpeter, J. (1939). *Business cycles: A theoretical, historical and statistical analysis of the capitalist process* (1st ed.). New York: McGraw-Hill.
Smith, A. (1789). *An inquiry into the nature and causes of the wealth of nations* (5th ed.). London: Strahan and Cadell.
Van de Ven, A. H. (1986). Central problems in the management of innovation. *Management Science*, 32, 590–607.
Van de Ven, A. H., Angle, H. L., & Poole, M. S. (2000). *Research on the management of innovation: The Minnesota studies.* Oxford: Oxford University Press.
Zaltman, G., Duncan, R., & Holbeck, J. (1973). *Innovation and organizations.* New York: Wiley.

Part I

Problematizing Innovation

2 On the Limits of What Can Be Said about 'Innovation'

Interplay and Contrasts Between Academic and Policy Discourses

Martin Fougère and Nancy Harding

This chapter engages in a Foucauldian genealogical trek (Foucault, 1980) to explore how innovation has come to be understood as something that is both desirable and measurable, so that it is impossible to talk or think about innovation outside these terms. A genealogical trek involves tracking back through time to interrogate the language used when writing about a topic and the changes in the language, what has become speakable, over the course of an epoch. This allows us to understand what limitations are put on our ways of thinking about that topic through the restricted language available for speech and, the opposite, what possibilities for thought are opened up by the changes that evolve in the language available for use. In this chapter, we focus on two influential ways of thinking about innovation that emerged in the 1960s. The first of these is a sociological approach, inaugurated in Rogers's (1962) first edition of *Diffusion of Innovations.* The second, the U.S. Department of Commerce's (1967) Charpie Report, is a policy text that inserted innovation into a logic of managerialism. Both of these texts are considered to have been seminal in the articulation of 'innovation,' with highly contrasting styles despite their common objective to understand innovation with policy considerations in mind. Five editions of Rogers's text have been published (1962; 1971; 1983; 1995; 2003), and the Charpie report has been followed by the *Oslo Manuals* (Organisation for Economic Co-operation and Development, 1992; 1997; 2005), enabling a genealogical exploration of the parameters within which innovation can be defined and understood.

Rogers's sociological approach puts people at the heart of innovation studies, whereas the managerial approach ignores people and favors a science devoid of any human element. The managerial approach has proved to have a much more sustained influence than the sociological approach. In this chapter, we provide some ideas why the managerial approach has become so much more influential today—and how the sociological approach may ironically have contributed to its own demise.

This chapter is thus in line with Godin's (2008) call for a genealogy of the category 'innovation.' However, we understand this call as pointing to a need to denaturalize innovation (i.e., to problematize what seems natural and universal about its dominant understandings) and to try to understand its peculiar power to entrance us. We thus contribute to Godin's (2008; 2010a; 2010b) historical studies on the development of the concept of innovation through exploring what is and is not sayable about the topic.

We turn next to summarizing the methodology used, and then begin our analysis by tracking Rogers's sociological approach. We follow this by examining the Charpie Report and the *Oslo Manuals*. We end the chapter with a conclusion that reflects on the juxtaposition of and possible interplay between the academic and policy texts studied.

METHODOLOGICAL APPROACH

The texts we explore here should not be seen solely as different ways of thinking about innovation, but rather as 'sites of definitional creation, violence, and rupture in relation to particular readers, particular institutional circumstances' (Sedgwick, 1991, p. 3), whereby their performative aspects, or their capacity for bringing into being that of which they speak, must be explored. They are a link between the cultural and social worlds, and they allow us to understand the processes by which we understand 'innovation.' The question arises of how to analyze the texts we are focusing on. Sedgwick's (1991) book follows a tradition of close analysis of influential texts to understand the present. Such a method of analysis has, in the last 30 years, been greatly influenced by Foucault, Laclau and Mouffe, and Derrida.

Foucault (1980) and Laclau and Mouffe (1985) encourage us to explore the discourses that inform understanding and thereby have power effects on subjects, whereas Derrida in *Archive Fever* (1995) enjoins us to analyze the 'power of consignation,' by which he means not only consignation or 'the act of assigning residence or of entrusting so as to put into reserve (to consign, to deposit), in a place and on a substrate' but also 'the act of *con*-signing through *gathering together signs*' (p. 3). The documents we archive occupy their place by virtue of a privileged topology, a place intersected by law and authority, and it behooves us to analyze how the archontic, or patriarchal principle, becomes instituted, and how legalities or legitimacies depend on it. In this reading, texts become a house or museum in which are *con*signed knowledges, practices and institutions which 'transit . . . through an archive, the science of which is not at a standstill' (p. 35). What is consigned, in other words, is what is kept available for memory and thus works in and on the world.

Thus we must go beyond an analysis of rhetoric to explore how what is sayable came to be sayable. In Derrida's (1995) terms, we must explore the institution of the archontic, in Foucault's (1980) and Laclau's and Mouffe's (1985) we must analyze the text as a site of fluctuating discourses into which the innovator may enter. We must discover:

> the politics of the scientific . . . statement, where knowing what external power imposes itself on science . . . is not the main question, but what effects of power circulate among scientific . . . statements, what constitutes, as it were, their internal regime of power, and how and why at certain moments that regime undergoes a global modification. (Foucault, 1980, pp. 112–113)

Thus we must explore the performative nature of the texts—understanding 'performatives' as 'statements that, in the uttering, also perform a certain action and exercise a binding power' (Butler, 1993, p. 225)—and how discursive articulations compete for dominance in a given 'field of discursivity' (Laclau and Mouffe, 1985). For example, if and when an articulation of 'innovation' comes to dominate the multidisciplinary field of discursivity formed by 'innovation studies' (i.e., all those works meant to study 'innovation'), then this articulation becomes 'hegemonic' (Laclau and Mouffe, 1985) and naturalized. It is our contention in this chapter that the understanding of innovation articulated in the *Oslo Manuals* has become so dominant that it is close to such a hegemonic status. Hence, by juxtaposing this understanding with academic sociological discourse on innovation, we aim to denaturalize what has come to be understood as 'innovation.' Through close analysis of the chosen texts and how they have evolved over four decades—with a particular focus on how/whether they problematize the consequences of innovations—we can illuminate how they have contributed to bring about what we understand as 'innovation' today.

ACADEMIC SOCIOLOGICAL DISCOURSE: ROGERS'S PROBLEMATIZATION OF INNOVATION IN SOCIETY

Prefaces and introductory chapters create a point of entry to the 'world' of texts, as they select and arrange participants, events, and circumstances (Stillar, 1998, p. 28). They: show what has been thought significant at particular times; demonstrate evolution in thinking over time; illuminate the author's stance toward the phenomenon and; allow analysis of discourses and knowledge circulating at points in time that may have changed or disappeared at later dates. The prefaces and introductory chapters to the successive editions of Rogers's (1962; 1971; 1983; 1995; 2003) book on the diffusion of innovations therefore allow insights into what has become—and what has ceased to be—sayable over four decades. Only the first edition contains a chapter explicitly titled 'Introduction' (chapter 1); in all the other editions chapter one is titled 'Elements of Diffusion,' but it still serves as an introduction. In the first edition of the book, there are no specific chapters or sections devoted to consequences of innovation, but in all later editions the last chapter (chapter 11) is titled 'Consequences of Innovations.' A close reading of these chapters also serves as a basis for the present analysis of the evolution of the discourse.

When looking at the successive editions from a broad perspective, what is immediately striking is that there are many significant changes related to consequences of innovations from the first to the third edition, but then the changes related to these issues from the third to the fifth edition are quite minimal. The changes between the earlier editions can be considered to be 'important discursive interventions' insofar as they have the power to frame 'innovation' in new ways over time, but the relative absence of change in later editions is also informative, as it suggests that Rogers's definition of the field was now established. The first three editions should therefore be seen as successive attempts at defining the field of inquiry so that it includes both the concept of 'innovation' and the notion of 'diffusion.' Such definition sets the scene in which other researchers will act, given the increasing importance of this field of practice and inquiry between the 1960s and the 1980s.

The most important discursive interventions discussed here are: (1) the change of framing in the titles of the successive editions and what this signals, (2) the evolution of the definition of 'innovation,' (3) the increasingly important emphasis on 'diffusion' rather than 'innovation,' and (4) the addition and development of a discussion (especially the last chapter) specifically devoted to consequences of innovations.

Changes of Framing in Titles

The first striking change between the first and second edition is the change in the title from *Diffusion of Innovations* (1962) to *Communication of Innovations: A Cross-Cultural Approach* (1971). In the preface to the second edition, this is partly explained by the fact that the first edition was actually—quoting the last words of the first edition—'the first of two volumes' to be complemented by a second volume 'ten or fifteen years' later 'after the leads for research suggested [in the first volume] have been followed up and expanded upon' (1962, p. 316; 1971, p. xvii). Rather than just an extended version of the first edition, the second edition—even though acknowledged as 'similar' to the first (1971, p. xviii)—is also a 'new intellectual venture' (1983, p. xv; 1995, p. xv), an expression used by Rogers in the third and fourth editions despite their greater similarity to their previous editions.

The two main additions that Rogers lifts up in the preface to the second edition—and that are signaled in the changed title—are: (1) more interest in and evidence as to whether 'the diffusion models of 1962 are not entirely culture-bound' (1971, p. xvii), leading to an 'attempt to highlight cross-cultural similarities, as well as differences, in the diffusion of innovations' (p. xviii); and (2) much more integration of 'diffusion research with the scientific study of human communication' (p. xviii). When we come to discuss the Charpie and Oslo reports, we will see that such richness of context is absent, with cross-cultural differences ignored in the former and in the latter regarded as a challenge to be overcome if the Organisation for Economic Co-operation and Development (OECD) nations are to have data that can be used to make comparisons. The word 'human' does not feature in these reports, in striking contrast with Rogers's texts, as we will show. Interestingly, even though more attention is given to consequences of innovation in Rogers's second edition, this is not explicitly articulated as an important development in the preface.

The change of title in the second edition (*Communication of Innovations: A Cross-Cultural Approach*, 1971) was to last only for that edition, with *Diffusion of Innovations* serving as the title in all other editions. The preface to the third edition explains that this title best exemplifies the research field studied in these books. The additional argument that Rogers makes in this preface as a justification for reverting to the original title is that this new change helps in 'avoid[ing] confusion of these two volumes [the second and third edition]' (1983, p. xvi). One key justification for releasing new editions has been the growth of this field in terms of sheer amount of available publications (from 405 in 1962 to about 1,500 in 1971 and 3,085 in 1983). Ostensibly there has been no major redirection of the focus of the book with the new editions, but rather efforts to cover a fast expanding field. However, discourse analysis refuses such an easy conclusion. If we look closer at the experiment with the new title in 1971, and contrast it with the preferred title, we see something that our analysis will later suggest is significant. 'Diffusion' refers to the spreading out from a point or source in a sort of organic way. 'Communication' is more agentive: It necessitates a transmitter and receiver, which, as the subtitle 'A Cross-Cultural Approach' suggests, would be communication across cultures. The first implies a dominant source of ideas that are then spread; the second implies a sharing of ideas and learning from each other.

Evolution of the Definition and Conceptualization of 'Innovation'

Rogers's definition of 'innovation' is introduced for the first time on page 13 of the 1962 edition, within chapter 1, which is titled 'Introduction.' The definition is short and remarkably open: 'An *innovation* is an idea perceived as new by the individual.' This must be contrasted with the Charpie/Oslo reports, in which the individual does not feature at all. There is a 'subject' in the *Oslo Manuals,* but as we will see, this is not a human subject. The Charpie Report relates innovation directly to research: innovation is what should happen after an invention or discovery, but there is no evidence in the report of any human involvement.

In all editions of the book, Rogers has been faithful to this understanding of innovation—with small additions over time that do not make the definition more specified, but if anything make it even broader (additions in bold):

> An *innovation* is an idea, **practice, or object** perceived as new by **an** individual (1971, p. 19)
>
> An *innovation* is an idea, practice, or object **that is** perceived as new by an individual **or other unit of adoption** (1983, p. 11; 1995, p. 11; 2003, p. 12)

Rogers thus puts the subject at the centre of its universe. This could be understood as pointing to the inherently social nature of the innovation phenomenon. One interpretation of Rogers's articulation is that it is the 'embedded self' (Townley, 2008, p. 92) who determines what is an innovation and also whether this innovation has undesirable consequences. Rogers's epistemology can thus be seen as not objectivist but relying on the historical social identity of the individual. We can interpret Rogers as drawing on an 'embedded rationality' (Townley, 2008), which is in sharp contrast with the highly disembedded policy texts that always need to refer to knowledge characterized by 'value-neutral objectivity and impartiality' (Townley, 2008, p. 92).

However, we can also see that Rogers relies on a very modernist notion of the human, that is, someone who looks out at (perceives) an objective world from her or his particular vantage point. In modernity, the visual is about nothing more than a perception that is pure and utterly descriptive of what can be seen. This 'sustained visual constraint of the [scientific] modern era' (Jenks, 1995, p. 7), described an ideology of seeing within an 'ocularcentric' culture, where cognition was to be informed through experience—only that which could be seen with one's own eyes could be believed, and what could be seen was beyond interpretation or misapprehension. The subject gazed out at an object that was there, pristine and untouched, waiting passively to be looked at and apprehended as it really is (Slater, 1995). The postmodern critique of this is that the visual has no epistemological certainty—the person who does the looking does not apprehend something that is given and objective, but interprets that which is seen. The object, meanwhile, 'looks back' and becomes embedded in the self of the viewer. The focus thus shifts to include not only the object that is represented but also the subjects engaged in the process of representing (Fuery & Fuery, 2003). The visual and, more broadly, perception, are now seen as political, for representations are a way of controlling and working on individuals. Prendergast (2000), indeed, argues that representations cannot be understood without incorporating an analysis of power.

So when Rogers goes on to argue the importance of the newness of the innovation to the individual—'it is the newness of the idea to the individual that determines his [sic] reaction to it' (1962, p. 13)—he should be interpreted as placing the (mythical) individual at the centre of the universe but devoid of any context. We do not know what power will work on the individual to influence her or his interpretation of what they define as innovative; we are ignorant of their education, their culture, or numerous other ways in which they are brought to interpret what they perceive. This shows the limitations of Rogers's definition: there can be no pristine individual interpreting innocently and untouched by power the degree of innovativeness of an object presented to them. However, this is not to deny that the social consequences of innovations are implicitly (and sometimes explicitly) among Rogers's main interests from the outset.

The other strikingly 'open' aspect of this definition lies in referring to innovation as 'any new idea': Rogers is interested in all kinds of social innovations (for instance, 'social movements' or 'clothing fads'), but again despite the inherently social nature of innovation that the definition entails, there is no explicit emphasis on context, culture, or power. This is surprising to the 21st century reader, because on the very first page of the first edition, Rogers writes about innovations as 'ideas that diffused into [people's] daily lives and emerged as cultural changes,' which suggests on his part a broad interest in social and cultural consequences of innovations. For instance, he claims that 'the explosion and shock of innovations on the minds of people are just as real and devastating as the nuclear missiles of modern military machines' (1962, p. 1), an analogy with war and the destructive side of innovation that resonates with the Schumpeterian articulation of 'creative destruction,' although what is in focus here is the impact on society rather than on the market. Very early on, there is an explicit problematization of possibly undesirable consequences of innovations: 'it should not be assumed that the diffusion and adoption of all innovations (for instance, a new narcotic among drug addicts) is necessarily desirable' (p. 3). However, there is no analysis of who has the power to define what is a 'problem' and what is or could be normative.

We turn now to a case study used by Rogers in all editions of his text. The fact that Rogers has chosen to keep using this illustration to start the successive editions (1971, p. 2; 1983, p. 1; 1995, p. 1; 2003, p. 1) suggests that the very socially and culturally grounded understanding of this case study is of key importance to his endeavour. To Rogers, innovations and their diffusion can only be understood in their social context: as Katz (1961, p. 16) put it, 'it is about as unthinkable to study diffusion without some knowledge of the social structures in which potential adopters are located as it is to study blood circulation without adequate knowledge of the structure of veins and arteries.' The case study in question is focused on 'innovation that failed: water boiling in a Peruvian town' (p. 7). Rogers chooses to illustrate the failure of an innovation by focusing on a public health service innovation that is neither commercial nor technological, unlike the most influential texts of innovation policy discourse. The conclusions on the case (after an anthropological study of 'the people in Los Molinos,' pp. 8–12) are expressed as follows: 'The reasons for the failure of the water-boiling campaign can be traced largely to the cultural beliefs of the Los Molinos people. . . . Only the least integrated individuals could afford to defy the community norm on water-boiling.' It is clear from this account and how it is presented that the innovation would disrupt social order, and in the first place people's positions in society would be threatened in such a way that only those who have little to lose can afford adoption.

On the one hand, this can be seen as a clear problematization of the consequences of innovations in terms of the disruption and reorganization of social relations and cultural norms that they may lead to. This focus on potentially undesirable consequences of innovations on society sharply contrasts with the stance in the Charpie/Oslo texts as we will see. However, Rogers's emphasis on understanding context does not extend to understanding the social context in which he writes. We must ask: why did he choose an example of a less-developed economy to illustrate his case? Is this not an example of the type of writing critiqued by postcolonial scholars, who argue that the exotic 'other' is used to prove to 'the West'[1] its own superiority—not just by describing negative features of the non-West, but also by romanticizing it (see, e.g., Said, 1978)? Rather than being seen as a straightforward description of an innovation that failed, Rogers's text can also be read as showing that an inferior 'other' fails to grasp the importance of something that would clearly benefit them if only they had the required qualities to grasp it and accept the change it promises.

Here we see the first coupling of 'failed innovation' with the less-developed other. We will go on to argue that successful innovations are part of the constructions through which the West knows itself as 'modern' or 'advanced.' Innovations therefore cannot fail, because failed innovations would have devastating ontological consequences—we would no longer be 'advanced.'

Increasingly Important Emphasis on 'Diffusion'

The focus of Rogers's first edition is almost as much on 'innovation' as a social phenomenon per se as it is on 'diffusion' as what leads to social change. But in subsequent editions, slight alterations in the articulation of the problem clearly make 'diffusion' the more central notion. In the second edition, the case of 'Water-Boiling in a Peruvian Village' (1971, p. 2) is still framed as '*An Example of Innovation That Failed.*' So it is still 'innovation' that is blamed for the failure. However, a difference in framing compared to the first edition is immediately clear from the title of the first chapter—which changes from '*Introduction*' (1962, p. 1) to '*Elements of Diffusion:* AN OVERVIEW' (1971, p. 1)—and from the very first sentence: 'This is a book about *communication*, a special type of communication, the diffusion of new ideas, new practices—innovations' (p. 1). The reframing so visible in the new title is a way to take the emphasis away from innovation toward communication and more specifically diffusion—even though, ironically enough, the term disappears from the title. The reframing becomes more complete in the third edition, in which the initial illustration of 'Water-Boiling in a Peruvian Village' becomes subtitled 'A Diffusion Campaign That Failed' (1983, p. 1). This slight change may be revealing of a more significant discursive shift: innovation per se is not the problem, rather what has failed has been the 'diffusion campaign.'

This shift can be seen to signify a (perhaps unconscious) less critical stance toward the potential negative impact that can inherently characterize an innovation. The threat and opportunity of change that could be seen to be embedded in the innovation (if we consider that innovation, by definition, is 'an idea, practice, or object that is perceived as new by an individual,' thus something that per se potentially leads to a change in how this individual conceives of, relates to, and acts in the world) is now attributed to whether diffusion succeeds or fails. On the other hand, this stronger emphasis on 'diffusion' further signals a sociological focus on the impact that innovations have through their diffusion, which confirms Rogers's interest in consequences of innovations. But as previously discussed, approaching these social phenomena from a 'diffusion' rather than a 'communication' perspective suggests a hierarchy, that is, a dominant (e.g., a developed Western) source of ideas that are then spread (e.g., to the less developed non-West), rather than a sharing of ideas and learning from each other.

Development of Discussion on Consequences of Innovations

The first time the construct 'consequences of innovations' is explicitly discussed is in the first chapter of the second edition (1971, pp. 17–18), although, as previously discussed, consequences were implicitly a central concern from the beginning of the first edition. Consequences of innovations are presented as corresponding to the third stage in the innovation-related 'social change process, following invention and diffusion' (p. 17). Thus, strictly speaking, consequences are demarcated from diffusion, that is, from the main focus of the book: in Rogers's theory they come into play after diffusion has happened. This is why they are treated separately from the rest of the book (in each of the latest four editions), in the last chapter.

The three key 'classifications of consequences' that Rogers (1971, p. 17, pp. 330–335) introduces are '*functional* versus *dysfunctional* consequences, depending on whether the effects of an innovation in a social system are desirable or undesirable' (these are later called '*Desirable Versus Undesirable Consequences*,' 1983, p. 380; 1995, p. 412; 2003, p. 442), '*direct* versus *indirect* consequences, depending on whether the changes in a social system occur in immediate response to an innovation or as a result of the direct consequences of an innovation,' and '*manifest* versus *latent* consequences, depending on whether the changes are recognized and intended by the members of a social system or not' (these are later called '*Anticipated Versus Unanticipated Consequences*,' 1983, p. 387; 1995, p. 419; 2003, p. 448). This typology articulates the possibility that consequences may be, for instance, undesirable, indirect, and unintended. This potentially problematizes innovations and their diffusion in a significant way, as the social change that is bound to happen as a result of the diffusion of an innovation may well be undesirable no matter what the intention behind and/or the direct impact of an innovation are. However, it is *how* Rogers illustrates this that is of interest here.

The first illustration that Rogers includes in relation to consequences is 'the case of the steel ax introduced by missionaries to an Australian aborigine tribe' (pp. 17–18, 335–336):

> The change agents intended that the new tool should raise levels of living and material comfort for the tribe. But the new technology also led to breakdown of the family structure, the rise of prostitution, and "misuse" of the innovation itself. Change agents can often anticipate and predict the innovation's *form*. . . . and perhaps its *function*. . . . But seldom are change agents able to predict another aspect of an innovation's consequences, its *meaning*, the subjective perception of the innovation by the clients. (p. 18)

One way in which the obvious clash in meanings is conveyed is by naming the illustration on pages 335 and 336 (and in all subsequent editions) 'Steel Axes for Stone Age Aborigines,' emphasizing the radical difference between Steel and Stone (Ages), which clearly suggests a cultural—and even civilizational—change, resonating with a sort of colonial (civilizing) discourse. One key aspect of the narrative is that 'before the days of the steel ax, the stone ax was a symbol of masculinity and of respect for elders' (p. 335). Then missionaries distributed steel axes 'to men, women and children alike,' and because 'the young men were more likely to adopt the new tools than were the elders, who maintained a greater distrust for the missionaries' (p. 336), this led to significant disruptions of symbolic and religious meanings as well as status relations—a 'confusion of age and sex roles' due to elders becoming 'dependent upon women and younger men' (p. 336). In addition, the new need for using steel axes led men to start 'prostituting their daughters and wives in exchange for use of someone else's steel ax' (p. 336).

On the one hand, this example clearly shows that consequences of innovation may be unintended ('latent'), indirect, and undesirable ('dysfunctional'). But on the other, postcolonial hand, look at how the exotic 'other' is again represented. The missionaries can be entrusted with this dangerous tool, in contrast to 'Stone Age Aborigines,' whose entire culture is cut down by the steel axe. Not only are the tribes people shown to be so backward they lack proper tools, they are also shown to have dubious moral standards (prostituting daughters and wives) and a tentative culture that can be destroyed by just one instrument. So again the concept of the 'failed innovation/diffusion' is related to 'the primitive' in contrast to the sophisticated West.

Another interesting aspect of the chapter from the second edition devoted to consequences of innovations is the discussion of why consequences have not been more studied until 1971 (pp. 324–326)—a discussion that will be reproduced along the same lines in every subsequent edition (1983, pp. 375–379; 1995, pp. 409–412; 2003, pp. 440–442), which suggests that despite the attention brought up by Rogers on consequences of innovations since 1971, very few other authors have studied these issues more recently. Here Rogers argues that: (1) there has been a tacit assumption (a sort of bias) among researchers studying innovations and their diffusion that consequences of innovation decisions will be positive, which has meant (both before and after 1971) that for institutional sponsors of diffusion research, adoption as such can be likened to 'innovation success'; (2) the dominant methodological approach within studies of diffusion of innovations, relying on surveys, has been largely inappropriate—and here the insights provided by Rogers's illustrations from studies relying on participant observation strengthen the argument, as the reader can indeed understand that consequences of innovations may best be studied through ethnographic methodology; (3) somewhat related to the previous point, 'consequences are difficult to measure' (quantitative methods may not be best suited for their study) and 'clients of change are often not fully aware of all the consequences of their adoption of a given innovation' (p. 325). That this same diagnosis is reproduced in every subsequent edition shows that the priority in innovation and diffusion studies has not been to study the noneconomic consequences of innovations and the potential negative externalities caused by innovations.

The final chapter on consequences of innovations in the third edition contains a number of additions compared to the second edition. In particular, Rogers includes a subsection titled 'the erroneous assumption of separability,' which makes it clear that 'most innovations cause both desirable and undesirable consequences' (1983, p. 382). This discussion is illustrated by the example of the Amish in the U.S., who according to Rogers have understood 'the principle of inseparability,' which leads them to 'willingly forego the desired advantages of tractors and modern farm equipment . . . in order to avoid the undesirable consequences of increased dependence on non-Amish businesses . . . lessened farm labor requirements, and the pressure for larger-sized farms' (p. 384). Although this is a balanced account showing neither excessive praise for the Amish mode of living nor conventional pro-innovation critique of cultures deemed to be 'backward', it can also be interpreted that by choosing a white, Western 'tribe' to illustrate this point, Rogers is implying that only Westerners are capable of refusing innovation without suffering negative consequences. Thus, although Rogers does put the pro-growth (and largely pro-innovation) bias that permeates Western societies into perspective, he also reproduces the sense of Western superiority that his previous ambivalent ethnographic accounts constructed.

A further addition that confirms that Rogers is not chiefly interested in commercial and technological aspects of innovations relates to his concern with the lack of equality in the consequences of innovations, particularly when it comes to 'development programs' aimed at developing countries (pp. 392–394). Partly drawing on communication research, he identifies different types of 'gaps' in the diffusion of innovations (pp. 394–397) and then analyzes 'gap-widening consequences of the adoption of innovations' (pp. 398–400), which leads him to suggesting 'strategies for narrowing gaps' (pp. 403–408). Whereas this discussion contains potentially very useful insights for policy purposes, it appears to be quite at odds with the influential policy documents that we analyze in this chapter. Whereas he clearly is concerned with the lack of equality between developed and developing countries in the consequences of innovations, his own writing can be seen as contributing to the very thing that he is against: his arguments elevate the West as the superior to a racialized, inferior other that is incapable of adopting innovations. This has important consequences, for it contains truths about how we know ourselves as 'the West.'

Implications of the Sociological Approach

On the one hand, Rogers's sociological approach is an interesting and intriguing counter to the dry managerialism of the reports that we go on to study. Rogers offers insights and ideas that are vital if we are to understand the reception to innovation by citizens, information that is conspicuously absent from the reports we analyze next. On the other hand, Rogers's texts, which were to become increasingly marginalized, set up definitions of innovation as part of the identity through which the West knows itself. That is, the West is modern in contrast to the premodern cultures of the non-West. That the West can handle innovations successfully, whereas its other is destroyed through its ignorance toward and incapacity in handling innovations, is core to that Western self-concept.

We have seen that when discussing unwanted and/or undesirable consequences of innovations, Rogers mostly uses examples referring to cultures that have kept a very traditional livelihood, and in some cases ('Stone Age Aborigines,' Amish) have not 'innovated' at all—even making a constitutive identity foundation out of this, as in the case of the Amish. These communities can be considered to be the 'other' of innovation; not only do they not innovate, but innovations will *have to* have negative consequences on them, because they will be bound to disrupt their most important wealth: an ancestral culture providing a sense of security and meaning.

One relevant question here relates to the absence of examples of negative consequences of innovations affecting 'innovation man'—by which we mean the contemporary Western consumer driven by permanent needs of self-actualization. This absence suggests that innovation man may not need stability or ontological security. Yet is innovation man not an embedded self too? How about the consequences of such an embedded self embracing an ethos of permanent innovation and consumption of innovative products, services, and experiences? Does this not lead people to a permanent sense of uprootedness—a disembeddedness that removes all ontological security?

And is this not problematic? By not explicitly illustrating the possibility of unwanted and undesirable consequences of innovations on developed Westerners, Rogers may ironically contribute to making his own analysis irrelevant from a policy perspective: if failures of innovations only concern marginalized communities, the overwhelming Western belief in the inherent 'good' of innovation is confirmed, at least as far as Westerners and Western-inspired consumers in the rest of the world are concerned.

In sum, Rogers's texts have the effect of relating failures in innovation to the 'non-West.' When we analyze the Charpie Report and *Oslo Manuals,* we will be drawn back to this understanding of innovation as something that defines the West, with failure of innovation challenging Western ontological security.

EVOLUTION OF MANAGERIAL AND SCIENTIFIC DISCOURSES OF INNOVATION: THE CHARPIE REPORT AND *OSLO MANUALS*' CONSTRUCTION OF INNOVATION AS SOMETHING THAT CAN BE MEASURED

Rogers's sociological approach to innovation declined in importance as the 20th century progressed, and this coincided with the increasing influence of a much more instrumental approach that aims to measure and thus increase the amount of innovations taking place.

The Charpie Report

The Charpie Report (CR) (U.S. Department of Commerce, 1967) is widely regarded as being 'the first governmental survey of technological innovation *per se*' (Godin, 2008, p. 42). Yet the report used 'rule of thumb' measures of innovation, based on the personal experiences of the panel members, because of an absence of firm data that its authors lament throughout the report. For example, the authors explicitly stated that [their] analysis 'could not be based upon empirical data on the innovative process,' but that 'it would be highly desirable to encourage systematic studies of the innovative process in order to clarify the strategic elements which stimulate and further innovation' (p. 8). When the report makes its recommendations, its authors write:

> It would please us to be able to say that our evaluation was made on the basis of clear, statistical evidence of the prevalence and importance of a given barrier to innovation, or on the basis of a sophisticated cost-benefit study of the impact of a given tax change on the amount of innovation or even on the level of tax revenues. Unfortunately, there are few such data available. In fact, the lack of objective data, in or out of government, on the innovation process, in general, and the technologically

> based firm, in particular, is symptomatic of a very serious deficiency in our thinking regarding technological innovation. (U.S. Department of Commerce, 1967, p. 30)

Rather than being, as widely reported, the first governmental survey of technological innovation, the report is an influential voice calling for such surveys. However, it may be the first report which presumes that innovations *can be counted*. The report begins by making a distinction between invention ('to conceive' the idea) and innovation (to use that invention, that is, the 'process by which an invention or idea is translated into the economy,' [U.S. Department of Commerce, 1967, p. 2]). History shows that it thus 'conventionaliz[ed] a definition of innovation as technological innovation and as a process leading to commercial innovation' (Godin, 2008, p. 43). However, rather than 'solv[ing] one of the main methodological problems confronting science and technology statisticians—how to measure output' (Godin, 2008, p. 12), the report emphasizes that any figures it uses are 'rule of thumb,' that 'we lack adequate empirical data to substantiate . . . feeling' (U.S. Department of Commerce, 1967, p. 15).

This leads to the question of why the Charpie Report should be seen as being the first attempt to measure innovation—what does this birth myth signify? We suggest, first, that Derrida's (1995) discussion of 'the archive' offers an answer to this question. That is, the very amateur, rule-of-thumb measures that are used in the report could, if not literally written out of history, pollute the later belief that innovation is somehow capable of being captured through scientific processes of measurement. Second, the report served to legitimize management through elevating it beyond the organization to the level of the nation state. The manager, in the Charpie Report's perspective, is necessary for guaranteeing that USA Ltd. continues to be profitable, with subsequent developments, such as the *Oslo Manuals,* being designed to develop tools to make that managerial task more efficient. Alternative ways of thinking about innovation, such as Rogers's sociological perspective, are silenced because (1) they threaten the (seeming) rationality and logic of a managerialist vision, which promises security and certainty; whereas (2) they may be understood to potentially challenge Western economies' perceptions of themselves as 'advanced.'

The Charpie Report's focus is on how to improve the management of innovation. It sees the most important factor affecting underspending on innovation to be 'the absence of adequate managerial and technological skills in an industry' (U.S. Department of Commerce, 1967, p. 15). It uses research and development (R&D) percentages as proxies for innovation, acknowledging openly the limitations of this approach. It locates the main reason for inadequate innovation in any industry in a management that has 'not learned the importance of utilizing technological opportunities and innovative skills' (U.S. Department of Commerce, 1967, p. 16). The major barrier to innovation, it concludes, is 'one of attitude and environment' and is 'primarily a problem of *education*' (U.S. Department of Commerce, 1967, p. 16, emphasis in original).

As hinted at earlier, in many ways the Charpie Report imagines the U.S. as nothing more than a gigantic corporation whose interests lie in maximizing productivity and profit. For example, on page 4 it compares changes in U.S. Gross National Product between 1945 and 1965 with the growth of five major corporations, as if they are equivalent. This is perhaps not surprising if the composition of the panel that wrote the report is examined, because it is dominated by business people. Its 16 members (all men) comprised eight representatives of industry, including Robert Charpie himself, who was then president of Union Carbide Electronics; three were attorneys; three were academics (one each from the Sloan and Harvard business schools, and a professor of law); one was from the Brookings Institution, and one was from government. Their reliance on academic research is minor: in chapter 2 they cite five studies to show that the vast majority of inventions come from small companies (in citing this research they start, curiously, with the title of the authors, emphasizing the 'professor,' thus legitimizing this part of the analysis), but rely on their own experiences to develop many of the ideas in the report. Thus two of its five chapters are devoted to exploring innovation in, respectively, small and large companies, but hypothetical examples are used that, later in the report, become the proof on which its recommendations are based.

What we see in the report is a very familiar style: a management report couched in a managerial language of maximizing profits. Its main focus is on solving the problem of getting innovations to the 'market place'; it concludes that there is no need for more R&D, but rather there is a need to 'look increasingly at the innovative process the way businessmen do' (U.S. Department of Commerce, 1967, p. 31), that is, the whole process from invention to market. Of its 16 recommendations, only one, the ninth on the list, concerns the need for studies of the innovation process. However, it makes a further 'over-all' recommendation in chapter 6, a 'central proposal,' which is that:

> The major effort should be placed on getting more managers, executives, and other key individuals—both in and out of government—to *learn, feel, understand* and *appreciate* how technological innovation is spawned, nurtured, financed, and managed into new technological businesses that grow, provide jobs, and satisfy people. (U.S. Department of Commerce, 1967, p. 56, emphasis in original)

This 'central proposal' would entrench management at the core of innovation. Summing up, the authors 'find that the concepts, uncertainties, and other realities of technological innovation are like a foreign language, indeed a strange world, to too many of us. Because of this, [they] believe the most important initial task before [them] is to become more widely acquainted with the "language" and "world" of innovation' (U.S. Department of Commerce, 1967, p. 57). This would seem to be a long way from innovation policy, the focus of the *Oslo Manuals,* to which we now turn.

The Oslo Manuals

The *Oslo Manuals* (OMs) (Organisation for Economic Co-operation and Development, 1992; 1997; 2005) provide guidance for carrying out a survey of the state of innovation in an economy. The first one (OM1), published in 1992, is no longer available. The second (OM2) is available as a free download on the Internet, as is the third and most recent manual (OM3), which is also available in book form.

Innovation policy 'has developed as an amalgam of science and technology policy and industrial policy' (Organisation for Economic Co-operation and Development, 2005, p. 15), but they cover innovation 'in the business enterprise sector only,' and deal with 'innovation at the level of the firm' (2005, p. 16). This is why we can argue that, rather than the measurement of innovation (if such a thing were possible) being a branch of science, it is another tool in the manager's toolbox, albeit one legitimized through claims to scientific credibility. However, the manager's toolbox is now used to govern the state, because those who are being advised in these reports are policy makers. We therefore see in innovation policy the managerialization of the state, where politics shifts away from ideology and beliefs and toward the best ways of managing an economy. An innovation is something that must be 'new to the firm' (2005, p. 18), and is related, just as in the Charpie report, to bringing this new thing to the market. Innovation activities therefore include such activities as development for preproduction, production, and distribution, support activities such as training and market preparation, and so on (p. 18). So, when the *Oslo Manuals* emphasize that 'the information needs of policy makers and analysts are a paramount consideration' (1997, p. 6; 2005, p. 15), these 'needs' are managerialist needs (making the economy run more profitably), and so the government becomes more firmly entrenched as the branch of management responsible for managing a country or state.

The disappearance of OM1 provides us with a birth myth to complement that of the Charpie Report—it suggests that the very innovativeness of that first manual is now something to be forgotten about, as if the passage of time has rendered its arguments quaint, from another age perhaps. There are many differences between OM2 and OM3, the most significant of which is a change in title. OM2 has the title '*The Measurement of Scientific and Technological Activities,*' and the subtitle '*Proposed Guidelines for Collecting and Interpreting Technological Innovation Data.*' The words 'Oslo Manual' are printed in very large type beneath the subtitle. By 2005, the subtitle has become the title, and the word 'proposed' has been omitted: OM3 is called '*Guidelines for Collecting and Interpreting*

Innovation Data'; and the title 'Oslo Manual' has moved from beneath the title to above it.

First, the removal of the original title signifies a move toward a more all-encompassing understanding of innovation, because OM3 includes guidance on measuring innovation in the service sector. The removal of the word 'proposed' signals that discussion is now not possible: the methodology has been fixed, is stabilized, and it now remains only for OECD member countries to implement the survey.

Second, the vast bulk of the *Oslo Manuals* are devoted to providing guidance in how to design and distribute survey-type instruments that should measure the amount of innovation taking place in a nation state. This gives the impression of precision and measurability: innovation can be identified; and once identified, it can be measured. However, its definition of innovation betrays that impression. The attempt to define innovation distinguishes between 'the subject approach' and 'the object approach' (1997, p. 11; 2005, p. 103). The definitions change radically between OM2 and OM3, as shown in Table 2.1.

In social scientific terms, the subject is that which has subjectivity and agency, and the object is that which does not. Where Rogers's work was concerned with individual subjects, there is no mention of the human at all in the Charpie Report or the *Oslo Manuals*. Rather, in OM2, the firm is defined as a subject, an entity that can behave and act, whereas the object is the innovation the (agentive) firm makes. This makes sense in social scientific terms. However, by OM3, the terms 'subject' and 'object' remain but now they do not make sense. Instead, we see a focus on expenditure, which appears to become a proxy for innovations themselves. Two types of expenditure can be measured: that which analyzes all innovative activity—whether successful, still in process, or unsuccessful; and that which focuses only on innovations that have been achieved. The 'object' is, by 2005, not

Table 2.1 Subject Approach and Object Approach in the Oslo Manuals

	OM2	OM3
Subject approach	'starts from the innovative behaviour and activities of the enterprise as a whole'	'covers expenditure for implemented, potential and abandoned innovation activities . . . it is a straightforward extension of traditional R&D measurement'
Object approach	'the number and characteristics of individual innovations'	'The sum reported comprises total expenditures on defined innovations, or on the main innovation(s) implemented during a given period.'

that which does not have subjectivity (the products that the organization makes), but a measure that is blind to anything but success (a pro-innovation bias). It is also notable that OM3 uses the word 'traditional' in relation to 'R&D measurement,' giving a history to the process that legitimizes at the same time as it obfuscates.

In the *Oslo Manuals* between 1997 and 2005, we see the emergence of what appears to be a proxy for measures of innovation: the finances expended. However, Derrida teaches us that the continuing use of the terms 'subject' and 'object' cannot be innocent: they perform work. What we are suggesting, therefore, is that by 2005, the location of failed or planned innovations in the subject signifies that failures are internalized (within the subject), causing psychological damage to that subject that cannot offload them (onto objects). In other words, we suggest that by 2005, we are seeing a fear of failure of innovation. The 2005 manual has to be put into the context of its day, which is a time when, although the European and U.S. economies were booming, the BRIC (Brazil, Russia, India, and China) economies were challenging Western economic dominance. Failed innovations therefore take on a different meaning: They waste finances that could otherwise have been used to sustain the competitiveness and dominance of the Western economies.

The *Oslo Manuals,* in this reading, seek to bring management to the control of nature. They achieve this quickly, because by OM3, they are able to reduce all innovation to numbers—numbers of innovations and expenditure incurred. The possibility of failure is removed: Anything that is unsuccessful is not defined as an innovation. From 2005, there can be no such thing as a failed innovation.

Implications of the Managerial Approach

In the Charpie Report we identified a 'birth myth' whereby the report is claimed to have introduced a rationality that is not found when reading the report rather than relying on second-hand testimonies. Rather, there is in the report an attempt at managerializing innovation—bringing it under the control of managers. The history of innovation has therefore been rewritten: a foundation in management has been passed off as a foundation in science. The *Oslo Manuals* continue the attempt to 'scientize' innovation, through developing measurements that claim to be able to capture and measure something that seems very difficult to define. This leads to the question of what, therefore, is being measured? Do the measures themselves create those things that they claim to measure, and/or is there something more profound at work here? We have suggested there is, and that Rogers's work illuminates what that might be: To innovate is to be modern. To be modern is to be Western.

Through relating these reports to our reading of Rogers's texts, we further develop the thesis that innovation is an aspect of Western identity: We know who we are because we can innovate. Innovation brings together management (the Charpie Report) and science (the *Oslo Manuals*) as ways of guaranteeing our ontological security.

CONCLUSION

When juxtaposing these academic and policy texts, the most striking first impression is one of contrast. Already in the 1960s, when Rogers's understanding was perhaps at its most influential,, his definition of innovation was considered problematic by economists who found it too subjective; this led economists and those policy makers keen to follow their prescriptions to try to come up with more clinical definitions of the term and so distinguish themselves sharply from Rogers (Godin, 2010a). Since then, although the policy articulation of 'innovation' has become more and more narrow, serving objectives of competitiveness through innovation systems that tend to consider all innovation as a force of progress per se, Rogers has kept problematizing innovations as not necessarily always good for the economy and society. He also has illustrated negative consequences of innovations with ethnographic accounts relying on an embedded rationality strikingly different from the disembedded rationalities that characterize policy texts. This may be why his works seemingly have slowly drifted away from the mainstream of innovation research and are now considered less relevant, especially in relation to research policy. Indeed, today the type of sociological studies conducted by Rogers does not constitute nor belong to any of the 'cognitive communities' considered dominant within 'innovation studies' as identified by Fagerberg and Verspagen (2009).

However, to be sidelined does not mean to lack influence. We have suggested that Rogers's work is very active and powerful in another way, that is, it provides the terms through which the West can define itself as 'modern,' in contrast to a premodern world that cannot cope with innovations. Indeed, to make conclusions about the phenomenon of innovation and its impacts, Rogers studied examples from less-developed (and 'less innovative') cultures, and so set up a dynamic of the developed West looking at the underdeveloped South in such a way that the West could know itself as modern through its difference with a 'backward' non-West (see Said, 1978) in terms of its capacity to innovate. The implicit message from 40 years of these sociological studies is that innovation (1) fails when it is attempted in less-developed cultures so that (2) by implication, innovation is something that is achievable only in the West. A dichotomy of modernity/premodernity is therefore established, in line with other cross-cultural analyzes in academic management discourses (see, e.g., Westwood, 2001; 2006; Fougère and Moulettes, 2007; 2012).

Despite Rogers's 'good anthropological intentions'—that is, striving to understand cultures in their own terms as much as possible—his examples, while romanticizing the other (Said, 1978), establish a relationship whereby innovation equates being advanced, and failure of innovation equates being backward/primitive. It is this that silences sociology, because 'innovation' becomes part of the definition of 'advanced' nations, and conversely to fail to innovate is to be 'backward.' The scientific/managerial approach fits into this presumption of modernity: It removes the unpredictable element (people) and so emphasizes science and rationality while disguising any impediment to that rationality. In other words, the West knows that it is modern because it can be innovative. To fail to innovate, therefore, is to challenge the ontological condition of the West.

In addition, Rogers's tendency to depict situations in which the innovative Westerners are those who bring innovations to underdeveloped communities suggests that, where attempts at societal improvement are concerned, 'we' have nothing to learn from these communities—this is in line with our observation that 'diffusion' significantly differs from (two-way) 'communication.' This is of course not surprising given the 'anthropological gaze' that characterizes several of Rogers's most prominent illustrations, but on the other hand, the understanding and even appreciation with which he describes the Amish, for example, would suggest that there may be something more 'human' in 'them' than in 'us.' By establishing an absolute opposition between the noninnovative communities and 'us' who try to help them, Rogers suggests that they do not necessarily benefit from our attempts to teach them, but he fails to see that we may need to learn from them.

This suggests that the competitiveness imperative may have been fully internalized already in 'our' case: Whereas 'backward' communities can be deeply studied as social systems free from the primacy of economic growth and progress, 'our' developed societies are constructed around those naturalized principles that are unquestionable (and made even less questionable by policy documents such as the Charpie Report and the *Oslo Manuals*), and thus 'we' are deprived of the possibility to learn from others but rather are condemned to the eternal loop of learning—from ourselves, always reproducing more of the same (just faster and faster) in the name of novelty. This never-ending innovation imperative cannot be problematized with the definition of innovation that is proposed in the policy texts. Indeed, if innovation is defined, as it is in the Charpie and Oslo texts, as changes that can be counted, and to be noninnovative is to be 'backward,' then that which fails as innovative, which cannot therefore be counted, is not innovation. In other words, it becomes impossible to speak of failed innovations—an innovation that fails is not an innovation.

When we turned to the Charpie Report and *Oslo Manuals,* we identified a turn to managerialism as a means of controlling innovation through counting it. Although the discourse of innovation seems to be moving toward closure around this dominant articulation centered on a commercial and technological understanding of innovations, that articulation itself disguises the work that innovation does in allowing the developed West to know itself. The need to count innovations right across developed economies, and to control this knowledge through the application of science used in the interests of a managerialization of innovation, circulates within this ontological positioning. Through managerialization of innovation, innovation becomes controllable and efforts can be made to continue its upward trajectory (in terms of numbers). The *Oslo Manuals* in particular therefore should be read as a means for ensuring that the West continues to know itself as modern, even, or especially, as its dominance is increasingly challenged by China, India, Brazil, Russia, and other 'emerging economies.' Having lost its empires, the West has long been able to reassure itself of its position in the world through its economic power; now that power is also being challenged. In such a case, what then is 'the West'? Who are we? If we can answer, 'we are that part of the world that leads the way in innovations', then we have an identity and we know who we (perhaps) are.

If we can acknowledge how fundamental innovation is to identity as well as to the economy, we can perhaps allow ourselves to open new and different ways of thinking about innovation, and in particular about 'bringing the human back into innovation studies.' A problematization of the consequences of innovation is needed: a critical rehabilitation of Rogers's socially grounded approach to an understanding of the impact of innovations in the social world might allow us to identify not only undesirable consequences of innovation, but also help us to better understand the changes that are made in its name and notably how these relate to the Western need for ontological security in the face of challenges posed by emerging economies.

NOTES

1. When we write about 'the West' in this chapter, it is to refer broadly, in line with postcolonial studies, to developed 'Europe-centered' countries, that is, mainly countries from Western Europe and 'Anglo-Saxon' countries, such as the U.S., Canada, Australia, and New Zealand.

REFERENCES

Butler, J. (1993). *Bodies that matter: On the discursive limits of 'sex.'* London: Routledge.

Derrida, J. (1995). *Archive fever: A Freudian impression* (E. Prenowitz, Trans.). Chicago: University of Chicago Press.

Fagerberg, J., & Verspagen, B. (2009). Innovation studies—The emerging structure of a new scientific field. *Research Policy,* 38, 218–233.

Foucault, M. (1980). *Power/knowledge* (C. Gordon, Ed.). New York: Harvester Wheatsheaf.

Fougère, M., & Moulettes, A. (2007). The construction of the modern West and the backward rest: Studying the discourse of Hofstede's culture's consequences. *Journal of Multicultural Discourses,* 2, 1–19.

Fougère, M., & Moulettes, A. (2012). Disclaimers, dichotomies and disappearances in international business textbooks: A postcolonial deconstruction. *Management Learning*, 43, 5–24.

Fuery, P., & Fuery, K. (2003). *Visual cultures and critical theory*. London: Arnold.

Godin, B. (2008). *Innovation: The history of a category* (Working Paper No. 1, Project on the Intellectual History of Innovation). Montreal: INRS.

Godin, B. (2010a). *"Innovation studies": The invention of a specialty (Part I)* (Working Paper No. 7, Project on the Intellectual History of Innovation). Montreal: INRS.

Godin, B. (2010b). *"Innovation Studies": The Invention of a Specialty (Part 2)* (Working Paper No. 8, Project on the Intellectual History of Innovation). Montreal: INRS.

Jenks, C. (1995). The centrality of the eye in Western culture: An introduction. In C. Jenks (Ed.), *Visual culture* (pp. 1–25). London: Routledge.

Katz, E. (1961). The social itinerary of technical change: Two studies on the diffusion of innovation. *Human Organization*, 20, 70–82.

Laclau, E., & Mouffe, C. (1985). *Hegemony and Socialist strategy: Towards a radical Democratic politics*. London: Verso.

Organisation for Economic Co-operation and Development. (1992). *Oslo manual*. Paris: Author.

Organisation for Economic Co-operation and Development. (1997). *Oslo manual* (2nd ed.). Paris: Author.

Organisation for Economic Co-operation and Development. (2005). *Oslo manual* (3rd ed.). Paris: Author.

Prendergast, C. (2000). *The triangle of representation*. New York: Columbia University Press.

Rogers, E. M. (1962). *Diffusion of innovations*. New York: Free Press.

Rogers, E. M. (1983). *Diffusion of innovations* (3rd ed.). New York: Free Press.

Rogers, E. M. (1995). *Diffusion of innovations* (4th ed.). New York: Free Press.

Rogers, E. M. (2003). *Diffusion of innovations* (5th ed.). New York: Free Press.

Rogers, E. M., with Shoemaker, F. F. (1971). *Communication of innovations: A cross-cultural approach* (2nd ed.). New York: Free Press.

Said, E. (1978). *Orientalism: Western conceptions of the Orient*. London: Routledge.

Sedgwick, E. K. (1991). *Epistemology of the closet*. New York: Harvester Wheatsheaf.

Slater, D. (1995). Photography and modern vision. In C. Jenks (Ed.), *Visual culture* (pp. 218–237). London: Routledge.

Stillar, G. F. (1998). *Analyzing everyday texts: Discourse, rhetoric and social perspectives*. Thousand Oaks, CA: Sage.

Townley, B. (2008). *Reason's neglect: Rationality and organizing*. Oxford: Oxford University Press.

U.S. Department of Commerce. (1967). *Technological innovation: Its environment and management*. Washington, DC: U.S. Government Printing Office. (Also known as the Charpie Report)

Westwood, R. I. (2001). Appropriating the Other in the discourses of comparative management. In R. I. Westwood and S. Linstead (Eds.), *The language of organization*. London: Sage.

Westwood, R. I. (2006). International business and management studies as an Orientalist discourse: A postcolonial critique. *Critical Perspectives on International Business*, 2, 91–113.

3 καινοτομία

An Old Word for a New World, or the De-Contestation of a Political and Contested Concept

Benoît Godin

Wherein It Is Documented that Innovation Is a Political Concept
Which Has Been Contested for 2,500 years,
Together With an Explanation on
How and Why the Concept Got De-Contested
Over the Twentieth Century.

Over the last 60 years, innovation has become the emblem of the modern society, a panacea for resolving many problems and a catchword. The quest for innovation is so strong that some who made no use of the concept until recently have discovered its "political" value and have begun using it. Biologists now speak about animal innovation (Reader & Laland, 2003). Others go so far as to suggest that drugs such as Ritalin and Adderall, used to treat psychiatric and neurological conditions, should be prescribed to the healthy as a "cognitive enhancement technology" to improve the innovative abilities of our species (Greely et al., 2008). Still others began talking of social innovation (Mulgan, 2007)—a not-so-new concept, however, because it existed in the early 19th century.

How did we get there? As a matter of fact, innovation has not always been highly valued. For over 2,500 years, innovation was essentially seen as negative. The innovator was a heretic, a revolutionary, a cheater. Certainly, over this period, "every single day brings something new, but [until the 17th century] the new is not fundamentally different from what has already happened," as Reinhart Koselleck put it (Koselleck, 2002, p. 161), and authorities were prone to controlling and limiting innovation. It was only in the mid-19th century and even more so in the 20th century that the representations of innovation changed. How did a concept that had been seen as pejorative for so long come to be a term of honor and a central category of Western thought?

This chapter documents the origins and development of innovation as a category.[1] It discusses the representations and uses of innovation over time and their development. The main thesis of this chapter is that innovation is a political, and essentially a contested concept. It got de-contested in the 20th century for "instrumental" reasons by those who had previously made it into a contested concept, governments. In the course of this development, innovation got a "restricted" meaning—technological and commercialized innovation—and came to be unanimously believed as beneficial. As a consequence, the critical study of innovation got increasingly marginal attention among researchers and policy makers, that is, "innovative ideologists."

THE INNOVATION CONCEPT AND ITS HISTORICAL MEANING(S)

Innovation as concept goes back to Greek and Roman antiquity. Novelty was relatively routine and accepted at the time, such as in fields that "give pleasure," as Aristotle put it, what we now call "the arts." However, this was not the case with innovation. The concept was pejorative. It owes its existence to a metaphorical use of a Greek term (καινοτομία), whose meaning was the opening of a mine. Such was the meaning of the term in Xenophon's *Ways and Means*, a work on 'political economy.' In the political writings on change and stability of constitutions, innovation came to mean introducing change into the established order (Plato, *Republic* and *Laws*; Aristotle, *Politics*; and Polybius, *Histories*). Thus understood, innovation was forbidden. To properly appreciate this meaning, keep in mind that innovation is distinct from novelty. Novation and innovation refer to introducing or bringing in some new thing that changes customs and the order of things in a nontrivial manner.[2]

Innovation as Forbidden

It is through religion and politics (intimately linked to politics for centuries) that the concept innovation entered common discourse.[3] This occurred slowly from the 1400s onward and reached a climax in the 1630s in England, leading to the first controversy on innovation, between Charles I, his protégé William Laud, Archbishop of Canterbury, and Puritans such as Henry Burton and William Prynne (Godin, 2010b). Burton accused the bishops of "innovating" in matter of church discipline and doctrine, and urged people "not to meddle with those that are given to change," an expression from Salomon's proverbs that, in the decades following Burton's use of it, would be widely repeated. In 17th-century England, documents by the hundreds made use of innovation to discuss religion, using the word as such. Over a hundred of these documents made use of innovation in their title, a way (or strategy) to emphasize a polemical idea and get a hearing.

During the Renaissance, the concept of innovation shared a place with heresy in religious discourses, particularly after the Reformation. It was precisely during the Reformation that the fate of the concept was determined for the centuries to follow. In 1548, Edward VI, King of England, issued a declaration *Against Those That Doeth Innouate.* A century later, Charles I, while explaining to his opponents why he had dissolved the Parliament, protested against Parliamentarians' innovations and proclaimed that he had never innovated himself. Even a king did not innovate.

In the 17th and 18th centuries, everyone was accused of innovating, and everyone denied innovating. Against accusations of innovation, people reply to the accuser: "YOU are the innovator." Two polemical uses of innovation were dominant.[4] One was naming or labeling someone pejoratively as an 'innovator.' The other was using the term to support an argument against change: making use of analogies with history or other countries where innovation have led to 'disasters,' or pointing to morality, namely the evil character of an innovator or the undesirable effects of innovation.

Then the concept came to be equated with political revolutions and revolutionaries (Godin, 2011). The model was, of course, the English political Revolution of 1649. After 1789, the emblematic example of violent political revolution was the French Revolution. To many, the Democrat or Republican is simply a revolutionary innovator who proceeds by "violent methods." The conclusion was direct: (every) innovation is necessarily sudden and violent. It is no surprise, then, that the term innovation was rarely used by Republican theorists and writers to make a case for the commonwealth in 17th-century England or the "revolutionaries" in the 18th-century France. As used (very occasionally) by these authors, it was in the then-traditional pejorative sense, and more often than not in historical writings or passages or while discussing religious issues—as would also be the case among philosophers of the Enlightenment and political writers in the 18th century. One has to look to other types of literature, such as anonymous pamphlets, to study innovation in the political thought of the period.

Next, it would be social reformers' turn to be accused of being innovators. In 1858, William Lucas Sargant published a study against those "infected with socialist doctrines" or "social innovators" as he called them—the French St. Simon, Fourier, and Proudhon, and the political economists including Adam Smith—to whom welfare is the solution to social problems rather than work (Sargant, 1858). Educational reformers would get no better treatment in the 18th century, their "extremes" being labeled "innovations" (Winslow, 1835).

The loading of innovation with pejorative meaning was sufficiently dominant for the concept not to be used for many centuries, except in the hands of critics or those who wanted to deny that they innovated. Such was the case in science, where the concept kept its political and religious sense for a while. Certainly natural philosophers and scientists of the 17th and 18th centuries developed a "new philosophy" and a "new method," experimental in nature, and this novelty was widely displayed in titles (see Thorndike, 1957). But the scientist claims that he does not innovate. The scientist keeps to the current values of society, "not meddling with Divinity, Metaphysics,

Moralls, Politiks, Grammar, Rhetoric, or Logic" as the motto of the Royal Society put it (Hunter, 1995). Such was Francis Bacon's view: The *Novum Organum* (1627) discussed "novelty" in scientific method openly, but the *Essays* held a different perspective on "innovation." Similarly, the one and only place where Thomas Sprat used the term innovation in his *History of the Royal Society* is to defend new experiments against political, social, or religious changes: "after all the *Innovation*, of which they [new Experiments] can be suspected, we find nothing will be indanger'd, but only the *Physics* of *Antiquity*" (Sprat, 1667, p. 328).

Certainly again, from the Renaissance to the 19th century, the new, the strange, and the curious was everywhere and was valued to many extents in different *milieus* (Thorpe, 1937; Thirsk, 1979; Daston & Park, 1998; Berg, 1999)—such was also the case during the Middle Ages. However, this was not the case with innovation, to the point that another vocabulary developed that hides innovative aims. "Renovation yes, but no innovation" is a recurring opposition of the time. "Reformation" and "renovation" improve on imperfect things, but slightly and slowly. On the opposite side, innovation changes the existing state of affairs completely and suddenly and "turns things upside down."

Such a situation would persist until the mid-19th century—and later: In the 20th century, the sociologist Robert Merton had suggested a definition of innovation as antisocial behavior.[5] Gradually, innovation gained new meanings. Innovation came to be used positively to talk about **novelty of any kind** and its 'productive' effects, rather than about changes in established customs. To this end, one more pejorative connotation had to be abandoned: that of novelty. To some, human's invention was accepted (when limited to specific domains), but to others, novelty is fancy, curiosity, contemplation, subtlety, private opinion, and fashion. Innovation came to refer to 'inventing' (in the sense of 'creativity') something new. However, this is not enough to characterize innovation—or to distinguish it from invention. Innovation also carried the idea of 'introducing' a 'useful' thing to the world. The history of innovation as a category in Western thought is the development of these two ideas, as discussed in the rest of this chapter.

Origin and Diffusion of the Innovation Concept

As previously mentioned, the Greeks had a specific word (of metaphoric origin) for innovation. However, among the ancient Roman writers, one finds only a couple of occurrences of *innovatione* and *innovare*, whose meaning is renewing (a return to the past), not innovating. *Renovare* (also in the sense of renewing) was more widespread. Precursors to or synonyms of innovation in ancient Rome are *novitas* (novelty) and *res nova* (or *nova res*), which stand for innovation as a substantive, and *novare* for the action of innovating. Declinations for these words are as numerous as those for the Greek καινοτομία.

Table 3.1 Origins of Innovation Related Words

The Word/Its Origin	France	England	Italy
Innovation	1297	1297	1364
Innovate	1315	1322	XIVth
Innovator	1500	1529	1527

The word innovation came into use in the late 13th and early 14th century (see table 3.1). It is a combination of *in* (into) and *novare*: introducing novelty. Early uses of the word exist from the 15th and 16th century in political matters—Machiavelli; Kings' calendar rolls, letters, and laws both in England (Richard II, Henry VIII) and France (François I). However, the word was used only occasionally prior to the Reformation. Then Catholics in England began to argue that Henry VIII, Edward VI, and Elizabeth had innovated. English Puritans adopted the same argument beginning in the mid-16th century. As a matter of fact, the word became so 'popular' that one may observe a 'linguistic inflation' among English writers, who often translated Ancient and Italian authors in their own words. For example, whereas F. Guicciardini's *History of Italy* (1568) contains only one occurrence of *innouare*, one finds dozens of occurrences in English translations, such as that of Geoffrey Fenton published in 1579.

As an indicator of the diffusion of the word, let us draw a statistics from old dictionaries. Whereas the word innovation appeared very rarely in dictionaries of the 17th century (in England, 9 out of 323 dictionaries), by the 18th century, nearly half the dictionaries published in English and French provided a definition of innovation (786 out of 1850). In these dictionaries, innovation means both introducing or bringing novelty and/or changing customs. As examples:

> Innovation: introduction de quelque nouveauté dans une coustume, dans un usage desja receu.
> Innover: introduire quelque nouveauté dans une coustume, dans un usage desja receu.
> Innovateur.[6] (*Dictionnaire de l'Académie française*, 1694)

> Innovation: change by the introduction of novelty.
> Innovate: to bring in something not known before; to change by introducing novelties.
> Innovator: an introductory of novelties; one that makes changes by introducing novelties. (Samuel Johnson, *A Dictionary of the English Language*, 1755)

The two meanings of innovation (changing customs and introducing novelty) are present in earlier and later dictionaries too. However, context determines uses, and such was the case during the innovation controversy in mid-17th-century England. In a context where references to antiquity or authority reigned, innovation was seen as a change in customs and thus pejorative. It was used in this sense by authorities in the 16th and 17th centuries in argument against the innovators. To others, namely the innovators who had to minimize the scope of their innovation in the face of criticisms, innovation meant renewing: a nonradical change, a return to the past. At the opposite or end of the spectrum of meanings, when the context is that of a disposition of people toward progress and the future, as during the modern era, innovation is seen as introducing novelty into the world for its "productive" (fruitful, beneficial) effects.

One meaning of innovation is absent from the previously mentioned dictionaries. In both dictionaries, innovation is defined as an action: introducing something new. However, innovation was also, since antiquity, used as a substantive: a novelty (new ideas, behaviors, and objects). One had to wait until the 19th century for such a meaning in dictionaries, as in the *Dictionnaire de la langue française* from Émile Littré (1872–1877): innovation is defined as both "action d'innover" and "résultat de cette action."

Innovation as novelty of any kind (rather than change in customs) is the meaning through which innovation turned positive, and this occurred after the French Revolution (Godin, 2012b). As a matter of fact, France is one of the countries where innovation started being used with a positive connotation in the everyday discourse. It had nothing to do with technology and the commercialization of technology.

As economists and many of us understand it today—the 'projector' was badly perceived over the 18th and early 19th century. Innovation came to be used rather to talk about a new era. Yet, like the English theorists of the Commonwealth in the 17th century, the French 'revolutionaries' made rare use, if ever, of the word innovation to discuss their innovation (the Revolution). It comes rather from post-Revolutionary France and the 19th century. People become conscious or aware of radical changes everywhere: in politics (political revolution), economics (industrial revolution), science (scientific revolution), and society (Marx and the coming social revolution).[7] Many writers, from the anonymous to the most famous, made use of innovation to name or talk about a new era and the productive effects of innovations. To a certain extent, such use was the case in the past too—and I emphasize this for all that follows in the following pages—but for a different reason: the "age of innovation" and the "spirit of innovation" were pejorative expressions. Now, this age is one of praise.

Between approximately 1750 and 1850 there occurred, as Koselleck suggested, a "shift in the conception of time and a reorientation towards the future" (*Sattelzeit*), "against which structural changes are perceived, evaluated and acted upon" (1977, 2004b; Richter, 1995, pp. 35–38). Innovation is part of this shift. Innovation began to be discussed in positive terms in every domain as a rupture, a break with the past, using the vocabulary of revolution (Godin, 2012b). As a matter of fact, revolution is no longer negative (Reichardt, 1997)—at least in many *milieus*. As Diderot put it in the French *Encyclopédie*, "Les révolutions sont nécessaires, il y en a toujours eu, et il y en aura toujours (Diderot, 1751: 400)."

Together with politics, three early uses of the word deserve mention because it is through these that innovation got a positive hearing—although negative uses continue to exist too. One is history (of both religion and politics). Writers started discussing past experiences of changes and revolutions as innovation, made analogies with their own age and pointed to the beneficial consequences. The second use is law. The French Revolution needed new instruments to become reality. Changes in law and legislation came to be named and discussed in terms of innovation. The third influential use of the word is science, and it is here that one finds titles with innovation in it-rather than isolated occurrences of the term as in history and law. Whereas the authors who contributed articles to the French *Encyclopédie* of 1751 conveyed a pejorative meaning of innovation and novators ("la secte des novateurs")—a meaning widely shared in 18th century France—, a French inventor published a *Dictionnaire chronologique et raisonné* in 17 volumes from 1822 to 1824 (Touchard-Lafosse, 1822–1824)—falsely attributed to a *Société de gens de letters*—followed by a Belgian dictionary in 1836 (Delepierre, 1836), both featuring innovation in their titles.

It would be erroneous to think that the use of innovation in science is intimately linked to or emerged because of a discourse on progress or the modern age. For example, despite a vocabulary on novelty and modernity, innovation had no place during the early ancients-moderns quarrel. Neither was progress talked of in terms of innovation among philosophers and scientists of the 18th century. Certainly a disposition toward progress contributed to experiencing novelty consciously, from the Enlightenment onward, and may have contributed to the changing values concerning innovation. Yet, more often than not, innovation appeared timidly alongside other terms, when it did, with no discussion. Novelty was a far more widespread term. Things would change some decades later.

20TH-CENTURY DEVELOPMENTS OF THE INNOVATION CONCEPT

By the middle of the 19th century and the early 20th century, the use of the concept exploded and permeated the scientific literature, in medicine, chemistry, engineering, and instrumentation most of all. One thing is certain: as titles of the time attest, to the scientists "innovation" was novelty in *methods*—not technological innovation (Godin, 2012b). This explains why innovation got into the practical arts and learning, namely those fields most in need of improvement according to the scientists: innovation as the introduction of the scientific method in useful knowledge.[8]

Ironically, most of the titles came not from science but from the useful arts. This representation occurred gradually over the 19th century, first in France. Germany would follow, whereas England would keep to the subversive use of innovation for a while, and scientists there made no use of the term until much later.

Why a new term in the vocabulary of scientists? If one keeps in mind the revolutionary connotation of the word innovation up to then, it may have been used to emphasize the profound (revolutionary) character of changes and novelties of the time. To the scientists, innovation is the introduction of something absolutely new (the science's method), which has never existed before, and which was unique for its effects or "fruits," to use Francis Bacon's word. Innovation as 'revolutionary novelty' is central to understanding the conceptual developments in the century that followed: innovation became a metaconcept that condensed a new experience of novelty into one word. Novelties are now experienced as radical or 'revolutionary' and permanent changes, they encompass more and more spheres of society and are considered useful. One needs a 'new' term: novelties become innovations. In turn, innovation becomes a catchword, as the expressions of the time attest ("itch of innovation," "plague of innovation," etc.).

Innovation as Creativity and Originality

Innovation, like invention, carries connotations of ingenuity and creativity, but it is not discussed in these terms among philosophers, natural scientists, and engineers.[9] One has to look elsewhere for systematic discussions of innovation. It is in the writings of social scientists that, from the beginning of the 20th century, innovation came to be explicitly theorized as **creativity**. Innovation was no longer understood merely as change, but as deliberate work of man's imagination—a direct import from Romanticism (and its 'theories' on genius and the productive or creative imagination) (Engell, 1981) and the late 19th-century theories of invention. Someone was said to be innovative to the extent that he 'generates' (invents) new ideas or new things—a synonym for inventiveness. This remains a common meaning of innovation. As a matter of fact, the two terms (invention and innovation) are often used interchangeably or one as a subclass of the other (e.g., technological invention as one type of innovation).

Creativity places the emphasis on originality: inventing something entirely new or doing something differently. This is the meaning through which innovation came into widespread use over the 20th century in science and industry (technology), but also in law, linguistics, arts, literature, and music. The mid-20th-century ideas concerning research or research and development (R&D) reinforced this meaning. Researchers were studied as the ideal type of creative individuals (together with artists) and research, defined as "creative" as opposed to routine work (as in the early editions of the Organisation for Economic Co-operation and Development, OECD, methodological manual, known as Frascati Manual), was posited as the source of (technological) innovation.

Among theorists of innovation, originality is discussed in two ways. First, like invention, innovation as creativity and originality is discussed in terms of "combination," from the first theorist Gabriel Tarde onward: combining previous ideas or things into new ones.[10] Combination is a 'psychological' category whose genealogy comes from the Scots' empirical associationism and was thereafter used in the philosophical and the literary criticism literature on imagination (Romanticism). To many writers of the 20th century, the combination of existing ideas to produce new ones defines what innovation is. Early 'psychological' theories of innovation came from the economic historian Abbott P. Usher (1929) and anthropologist Homer G. Barnett (1953). In the hands of some economists, such as J.A. Schumpeter, the entrepreneur (who combines the productive factors) became the creative innovator *par excellence* (Schumpeter, 1934; 1939).

Second, the most original innovations, in the sense of those having the greatest effect on society, were described, introducing the vocabulary of revolution into the analyses. In fact, revolutionary and widespread changes have characterized the meaning of innovation in the very first theories of innovation, as it did for the early uses of the term among French writers in the early 19th century. That people experience change everywhere finds expression in the first theory of innovation, that of the French sociologist Tarde, to whom innovation covers social organization, politics, law, culture, religion, language, industry and arts (Tarde, 1890). Anthropologists of the early 20th century had a similar and large meaning of innovation. That changes are discontinuous and revolutionary also finds expression in the economist Schumpeter's concept of economic change or development through ("combination" or) innovation as "revolutionary change" (Schumpeter, 1934, pp. 62–64; 1939, p. 226) and "creative destruction" (Schumpeter, 1942).

Typologies were consequently developed for classifying innovations as "revolutionary" *versus* incremental. Revolutionary innovations got the attention of most theorists and gave rise to hierarchies of innovations with hyperbolic terms (major *versus* minor innovations; great, important, fundamental, basic, and radical innovations *versus* supplemental, incremental). Consequently, theoretical controversies ensued among historians, anthropologists, and sociologists. To some, innovation (or rather of the invention behind an innovation) is evolutionary, not revolutionary, in character. The irony is that to 'evolutionists,' the innovations responsible for advancement of society or economic growth were revolutionary. To sociologist S. C. Gilfillan, the development of inventions is evolutionary (the combination of many small contributions) but it is "revolutionary" inventions (with an "s") that change civilization (Gilfillan, 1935). Similarly, Schumpeter has contrasted evolutionism (concerned with continuous change and equilibrium) to the "discontinuity" and "revolutionary" character of innovations (Schumpeter, 1934, p. 62f).

Innovation as Utility

Over the 20th century, innovation took on a more specialized meaning related to the *use* of new ideas, things, or behaviors, whatever their source. Already during the previous century, innovation in scientific methods carried the connotation of utility. The inventor and the inventor-entrepreneur (projector) had the idea (or rhetoric) of introducing something useful into the economy too—but without using the vocabulary of innovation. Today, social researchers place explicit emphasis on **utility** or **usefulness**. Innovation is theorized in terms of "introducing" or "adopting" some novelty into practice (into groups, cultures, firms, or governments).[11] This is the second modern meaning of innovation: the use of new methods or ways of doing things, as scientists had suggested in the previous century. This is the meaning used in the literature of the second half of the 20th century in sociology, management, politics, and neoclassical economics—in the latter case, it explains innovation as the introduction of new (methods or) processes in industrial production.

In the last 60 years or so, theorists have developed two variants of this meaning. One is innovation as a **commercialized** invention. This meaning refers mainly to technological inventions and has become the dominant understanding: bringing (selling) technological inventions to the market. Whereas until then innovation has been understood as action (doing something differently, using new methods, processes, or practices), it is here limited to a thing: a commercialized product.[12] To economic historian W. Rupert Maclaurin, the first to put it explicitly as such, "When an invention is introduced commercially as a new or improved product or process, it becomes an innovation" (Maclaurin, 1953, p. 105).

Limiting innovation to technology is, to some theorists such as anthropologist Barnett, a 'restricted' meaning of innovation (Barnett, 1953, p. 8). In fact, except for a few authors (Tarde) and a few disciplines (anthropology), such a restricted meaning is spontaneous among theorists. Certainly, some have regularly talked of social invention or innovation (W. F. Ogburn, S. Kuznets, C. Freeman), but in the end, they chose to concentrate on the study of technological innovation.

The other variant of innovation as introducing something useful is innovation as a **process** over time, from the generation of an idea to "appliedness," from invention to diffusion. Whereas innovation as a thing is discussed in the plural form and gave rise to countless exercises in counting the number of innovations in an industry or a country, here innovation is discussed in the singular. As the panel on technological innovation from the U.S. Department of Commerce put it in 1967: innovation is a "process by which an invention or idea is translated into the economy" (U.S. Department of Commerce, 1967, p. 2). Innovation as process emerged as a solution to the early 20th century controversy as to whether invention or diffusion comes first in explaining the development of societies or cultures (Godin, 2012c). It gave rise to many sequential theories or "models" of innovation and, after World War II, to the highly influential but controversial "linear model of innovation."

Three Meanings of Innovation
Substantive: novelties (new ideas, behaviors, objects)
Action: introducing (or bringing in) something new
Process: from invention to diffusion (adoption, commercialization)

Further Distinctions

Over the 20th century, innovation came to be distinguished from (and opposed to) invention in the sense that innovation is useful invention put into practice. Innovation is now discussed in terms of translation or "**application**." Economic thought is responsible for the distinct meanings of invention and innovation—although the opposition is quite old and took many forms over time: scholastics/humanism, speculation/operation, discovery/invention, and basic/applied science. With innovation, the opposition coalesced and crystallized into a single word. To Josiah Stamp, invention "has too mechanical a connotation" (Stamp, 1937, p. 5); "mechanical and scientific discovery, even in practical form, is not economic wealth until man has learnt to enjoy it in an economic sense" (Stamp, 1929, p. 120). To Schumpeter, "innovation is possible without anything we should identify as invention and invention does not necessarily induce innovation." Invention is an act of intellectual creativity and "is without importance to economic analysis" (Schumpeter, 1939, pp. 84–85), whereas innovation is an economic decision: a firm applying or adopting an invention.

Innovation also came to be understood as opposed to another concept: imitation. Like invention, imitation defined innovation for a while. During the Reformation, when people denied that they were innovating, they were simply pretending that they were going back to original and purer times. They were imitating the past, which time or people had corrupted. If by chance or as a last recourse people admitted innovating, the innovation was claimed to be imitation. Christopher Dow, an English divine involved in the controversy against Henry Burton in 1637, put it as follows, "I cannot but wonder with what face he can accuse any of these things of novelty, when there is not one of the things he names which hath not been used in the primitive and purest ages of the Church" (Dow, 1637, p. 114).

Then, as previously mentioned, originality (rather than imitation)—which also defines invention—came to be the criterion or yardstick for innovation. In the 20th-century literature on innovation, originality refers to being *first* in doing something differently.[13] However, being first is difficult to assess: It is relative and subjective. A man can be an innovator compared to his own past behavior or versus his own group, and yet not be an innovator 'internationally' (or across time). He simply imitates what already exists elsewhere. It is innovation to him subjectively, but not objectively. To many, innovation is objective innovation only. The rest is **imitation.** As the economist Theodore Levitt put it, "Strictly speaking, innovation occurs only when something is entirely new, having never been done before" (Levitt, 1966, p. 63). Innovation as the introduction or use of something that already exists, as it was previously understood in the social sciences, is imitation. Similarly, to economist Jacob Schmookler, "the first enterprise to make a given technical change is an *innovator.* Its action is *innovation.* Another enterprise making the same technical change later is presumably an *imitator,* and its action, *imitation*" (Schmookler, 1966, p. 2).

In the following decades, many economists adhered to such a representation. For example, when discussing firms' strategies, C. Freeman minimized and contrasted "the traditional strategy [use of invention as] essentially non-innovative, or insofar as it is innovative it is *restricted* [my italics] to the adoption of process innovations, generated elsewhere but available equally to all firms in the industry". To Freeman, innovation "excludes simple imitation or 'adoption' by imitators." (Freeman, 1974, p. 257).

The representation has not gone unchallenged. In a study conducted for the Science and Industry Committee of the British Association for the Advancement of Science in the late 1950s, C. F. Carter and B. R. Williams suggested that a firm "may be highly progressive [innovative] without showing much trace of originality [research]. It may simply copy what is done elsewhere. . . . It is nonsense to identify progressiveness with inventiveness" (Carter & Williams, 1957, p. 108). This was also Schumpeter's understanding: an innovation "need not necessarily have occurred in the industry under observation, which may only be applying, or benefiting from, an innovation that has occurred in another" (Schumpeter, 1939, p. 89, footnote 1).

The conceptual difficulties of the distinction between innovation and imitation led many economists to abandon the category 'innovation' in *lieu et place* of "technological change," whereas other researchers and statisticians created classifications to measure levels or degrees of innovation or innovativeness (Rogers, 1962): a first chronologically or internationally (objective), and a first compared to or within one's neighbors, group, or culture (subjective). As for imitation, the concept gave rise to "diffusion" studies (Tarde, 1890; Rogers, 1962; Mansfield, 1968)—with imitation used as term for a while, instead of diffusion—and to the study of lags between innovators and followers or imitators (often termed laggers).

DE-CONTESTATION AND LEGITIMIZATION OF THE INNOVATION CONCEPT

Innovation has become a basic concept in Koselleck's sense (Koselleck, 1996, p. 64). It crystallizes into a single word a whole semantic field or

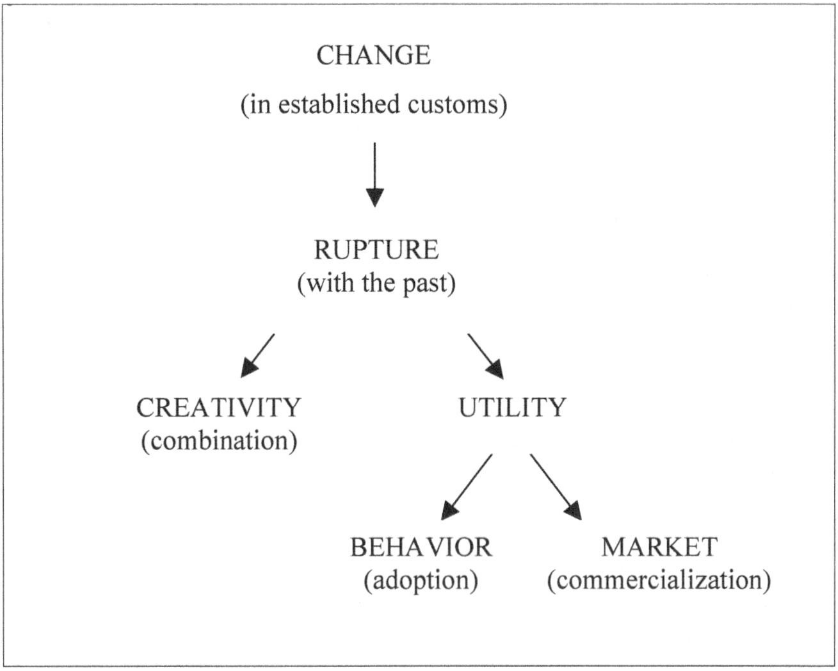

Figure 3.1 Representations of innovation over time.

cluster of other concepts and ideas: change, novelty, invention, creativity, originality, usefulness. (An overview is presented in figure 3.1.)

The 20th century has made of innovation an ideology, or a de-contested category: Innovation has acquired a dominant and (almost exclusively) positive connotation. De-contestation is a process of naturalization of contingent concepts, practices, and representations into ideologies (Norval, 2000; Freeden, 1996). As collective belief systems, ideologies de-contest political concepts by converting the variety of opinions into a monolithic certainty. They naturalize what is contingent and provide a particular organization of society, a nonhistorical, given, and natural articulation, an imaginary. Innovation as ideology precisely serves this function. Innovation has become a noncontroversial practice, an institutionalized signifier and an ordering and structuralizing principle of thought and action.

It is here that those who had made innovation a contested concept in the past (governments) de-contested the concept and made innovation a tool of public policy. In this task, governments have been supported by theoreticians. Governments themselves, together with social scientists as consultants, have acted as "innovative ideologists." They have brought forth a new vision of innovation (technological innovation) in response to a specific context (competitiveness) and in reaction to previous thoughts (pejorative), and have thus acted as advocates of innovation and sellers of expectations or promises. Following Quentin Skinner, I call these innovators "innovative ideologists" (Skinner, 2002a, chapters 8–10; 2002b, chapter 4; Godin, 2010a).

To governments and international organizations, innovation is technological innovation and is a tool of policy, a tool in the service of industry. The rationale is essentially economic: competitiveness between industries and between countries. In the late 1950s, a whole discourse developed in Europe about 'lags' and 'gaps' in science and technology between Europe and the U.S. To governments and international organizations such as the OECD, technological innovation became a means to economic growth, productivity, and market share, or competitiveness. Hence the need for policies to stimulate technological innovation.

The official representation of innovation also carried a second component: commercialized innovation. Governments have been among the first to systematically promote a definition of innovation as commercialized technological innovation, because of its relevance to policy issues (competitiveness and productivity). Until then, most economists, for example, had been working with the idea of "technological change" as the introduction of new inventions in industrial production, not their commercialization. Furthermore, as previously discussed, there were many definitions of innovation per se emerging in sociology, management, and political science. However, there was no accepted and standardized definition. Nonetheless, the OECD and governments selected one of these definitions (commercialization) with no hesitation. In the following decades, innovation as the commercialization of technological inventions became the dominant representation of innovation.

In this task, governments have been aided by social researchers who, as consultants, produced theories in coproduction with governments (Godin, 2012a). To legitimize innovation for policy purposes, governments needed arguments or conceptual frameworks. It was here that theoreticians entered the scene. The work of legitimization was conducted with the aid of social scientists who developed theoretical thoughts on technological innovation and the *factors* that firms and governments may influence or control to get more out of material or technological invention. Acting as consultants to the governments and international organizations, the social scientists acted as advocates (because sympathizers and beneficiaries) of innovation (whose source is said to be knowledge)—as much as 'scholars.' Then, social researchers contributed the theoretical thoughts that made innovation technological and commercialization, and served governments' ambitions and policies. They contributed to bringing technology's connotations—industrial and material—to innovation.

They helped to coalesce and crystallize the then-prevalent discourses on technology into a vision of innovation as technological.

FOUR MOMENTS IN THE HISTORICAL USES OF INNOVATION

Innovation is a concept we use unconsciously, often without knowing precisely its richness. Innovation does not exist in itself. It is constructed through the eyes and through discourse (Papon, 2004). This construction is the result of the contributions of many individuals over many centuries. Forbidden in the past, innovation has become an ideal everyone believes in. Today, everyone display his innovative performance.

From the Renaissance onward, innovation has been part of everyday discourse. Innovation haunts people. In the past, a person scarcely ever defined himself as innovator. No innovator thought of naming himself an innovator. However, one's own opponent was quickly labeled innovator. Over time, a term used only occasionally became a subject of regular thinking. Innovation gradually gained acceptance within a new social and political context. The uses of innovation developed according to four historical "moments":

1. Innovation as a **political** and essentially a contested concept. Before the 20th century, innovation had nothing to do with creativity and originality on the one hand and utility on the other. Innovating was seen as introducing change into the established order, and was explicitly forbidden by law.
2. Beginning in the 17th century, innovation turned into a social and **polemical** concept. It was used in pamphlets, tracts, and other documents aimed at opposing innovators of any kind, first of all in politics and religion, but also in law, science, and education. The term was a weapon against change. It was essentially perceived as negative.
3. From the 19th century, innovation turned into a positive category. It is tempting to think that technology is responsible for this change in attitude. However, before the industrial revolution, inventors had not yet become heroes (Macleod, 2007), and the inventor-entrepreneur or "projector" was no more trusted than he was before (and did not use the vocabulary of innovation). Certainly, there were some early tentative rehabilitations of the projectors (Defoe, 1702; Bentham, 1787; 1793–1795), but it would only be a century later that social researchers would succeed and make innovation a technological affair. Innovation became an **instrumental** category in the 1960s. Governments became convinced that technology is a source of economic progress. Innovation therefore acquired a dominant representation as technological innovation. Statistics have played a central role: Statistics gave identity and

objectivity to what had been a subjective concept. In fact, over the last 50 years, discussions of innovation have been intimately linked to measurement issues.

4. Innovation turned into a **theoretical** category in the footsteps of governments' thoughts and policy issues. Social researchers, and economists more than any other, developed theories to understand and explain (revolutionary) changes through innovation and how governments may help. By 1960 to 1970, the literature had exploded, giving rise to regularly updated bibliographies (Rogers, 1967), reviews (Kelly & Kranzberg, 1974; Radnor, Feller, & Rogers, 1977), handbooks (Dodgson & Rothwell, 1994; Stoneman, 1995; Shavinina, 2003; Fagerberg, Mowery, & Nelson, 2005) and research centers.

To these developments, one could add a fifth moment. From the 1980s onward, innovation became an end in itself: anything goes in the name of innovation; everyone should innovate. Innovation has become a slogan. To paraphrase Koselleck on revolution, innovation "is a widely used forceful expression whose lack of conceptual clarity is so marked that it can be defined as slogan" (Koselleck, 1969, p. 43). Discourses on innovation have become **performative**: they produce innovation in the sense that they encourage people to innovate and then reward them. Discourses on innovation create the world of innovation.

The history of innovation as a category is not a linear one (and the previously mentioned moments are additive rather than substitutive). Over a span of 2,500 years, innovation has remained a political concept: first as forbidden, then as a polemical discursive tool, then as an instrument of policy. Second, there existed positive uses of innovation before the 19th and 20th century, although they were few. Such is the case in the ancient literature on history (Livy) or Machiavelli in the early 16th century (Pocock, 1975): if the goal is useful (sometimes giving a voice to the people, at other times increasing the Prince's power) innovation is seen as praiseworthy. Livy gave an interesting discussion on innovation in Book 4 of his *History of Rome*. Discussing a law on the intermarriage of patricians and plebeians, he asked: "Ought no innovation be adopted?" (*Nullane res nova institui debet*?) just because it had never been done before. No. To Livy, the citizen had supreme authority and was "permitted, if it so desire, to enact a law."

The story is also not linear in two other senses. One finds (some) "theoretical" thoughts before the 20th century: Machiavelli and Francis Bacon discussed resistance to innovation and suggested (different) "strategies" to deal with the resistances. To Machiavelli, the Prince should innovate early and fast (to make people forget), whereas to Bacon one should innovate slowly, as time does, and not meddle with politics. Second, the evolution of representations was slow. In the 20th century, many social scientists began using innovation in their own field: in education, sociology, management, or political science. However, neoclassical economists remained reluctant to talk of innovation, except during a brief interlude on "induced innovation" in the early 1960s. They invented another concept in place of innovation, that of "technological change." That the concept innovation has become theoretical has thus not solved many problems. "What is innovation?" has remained a controversial question. In the last decade, official statisticians have decided not to use the term in questionnaires on innovation, following a suggestion first made in 1963 by the consultant Arthur D. Little.

Finally, innovation as a "fact" has remained a contested phenomenon over its whole history. Innovation and its value have varied according to people, fields, and contexts: literary writers offered resistance to innovation in the name of nostalgia for past ages, whereas others argued that changes threatened morality. Inventors-entrepreneurs or projectors in the 18th century were not trusted, Luddites in the 19th century opposed the introduction of machines in the factory, and people resisted technological innovation in the 20th century (Stern, 1937). Over the whole period considered in this chapter, there was always an essential ambivalence toward innovation. Novelty was to a certain extent accepted, whereas innovation was not. However, overall, the 20th century has developed representations of innovation completely different from the previous representations, in reaction perhaps to 2,500 years of "intellectual terrorism." Innovation is positive when it served a good cause, namely when it is useful. As an anonymous author put it in 1789: "On ne doit jamais craindre d'innover, quand le bien public est le résultat de l'innovation" [One should not fear to innovate when public welfare is the outcome] (Anonymous, 1789).

CONCLUSION

From antiquity onward, change (corruption, degeneration) was everywhere and was discussed by almost every writer, either to stabilize it or to bring about a complete revolution. Had change not been a focus of people's attention, there would have been no talk of innovation. In turn, the concept of innovation itself has changed considerably over time. To paraphrase Koselleck (on revolution) again, innovation "possesses such [innovative] power that it is constantly extending itself" (Koselleck, 1969, p. 44). In recent decades innovation gave rise to neologisms such as "innovativeness," "innovational," and "innovativity," an inflationary usage that every one of us can observe in the media, but which is present also in the scientific literature. Innovation has turned into a catchword or buzzword, with a legitimizing function.

In the last century innovation has become a widely accepted concept. Innovation got a positive hearing when people started experiencing changes everywhere, 'revolutionary' changes more than any other, and worked deliberately to make still more changes. No one individual has been responsible for the new representation. The economist Schumpeter is often credited for having discussed innovation early on, but he was not alone. It is only in retrospect that Schumpeter has become a spiritual and symbolic father.

As the concept of innovation has developed over time and transformed from negative connotations toward a positive and noncontroversial practice, the critical study aspects of innovation have been given neglected attention. I will conclude by summing up what has been discussed in this chapter, highlighting the main hypotheses—still in need of more validation.

What Is Innovation?

- Innovation is a **political** category. It was first discussed in political thought (and religion, the two being interwoven for many centuries), then forbidden (kings' declarations and proclamations) and regulated by law (patents, copyrights), then theorized about in terms of policy issues.
- Innovation is a **contested** category. Early on, it had been politically and religiously loaded and carried a negative connotation. Innovation was evil and subversive—the term was more often than not used together with other words (fancy, superstition, corruption, danger, violence, faction, rebellion, and dozens of negative epithets)—and innovators were pejoratively called "novellers," "novelists," and the like. Such a representation of innovation gave way to another vocabulary that talked about innovation in a positive sense (reformation, renovation; today we talk of modernization). When the term innovation was used, it was by opponents and critics against transgressors and deviants from norms, customs, and traditions. Later, namely when innovation gained a more positive value, resistance continued. Some preferred not to use the term or developed other terms in its place.
- Innovation is **change**: introducing change into established practices, then change as novelty in the sense of creativity and originality. The use of innovation together with change and alteration in the same linguistic expression is witness to this meaning. In the 20th century, innovation as change continued, as reflected in concepts such as culture change, social change, and "technological change."

Development of the Representations

- Innovation slowly turned **positive** from the mid-19th century onward, and became a term of honor over the 20th century. Through the centuries, innovation started off being a matter of fear, then became a matter of routine and satire, then became a cult. This happened first in post-revolutionary France (England lived for a while with subversive thoughts about innovation). From being a political (prohibition), then social (deviance) category, innovation became a historical category, theorized in term of rupture with tradition or traditional ways of doing things. Usefulness as opposed to mere invention and doing as opposed to thinking were what drove the transformation of innovation's representations.

- Beginning in the 1960s, innovation became **legitimized** by governments, working in coproduction with social scientists as consultants, particularly (nonmainstream) economists. The legitimization arose due to the role of technological innovation in industrial progress and the competition between countries.
- From the 1970s onward, innovation as **technological (and commercialized) innovation** became dominant in discourses, helped by statistics that crystallized this representation. Such a representation rapidly became spontaneous. Many use innovation without the determinative "technological" when talking of technological innovation, either deliberately, or more often, unconsciously. Few writers discuss what innovation is, but they make use of a disciplinary or "restricted" definition from the start (technological innovation) and make a pope of their preferred author (Schumpeter). We have come a long way from the 17th century, when innovation appeared in dictionaries of "hard" and "difficult" words (Phillips, 1658; Blount, 1661; Coles, 1677). To some, this has led to a loss of the variety and richness of the term, which many are currently trying to recover (social innovation is an example)—for reasons having to do with "getting a share of the attention" devoted to technological innovation as much as for purely intellectual reasons—and may (possibly) destabilize the dominant representation.

Until the last century, innovation was not used to discuss novelty and the like, but used pejoratively. In this chapter, we have asked why innovation came back into our everyday vocabulary with a positive meaning. We looked for an answer in the word innovation itself and have found that the issue, although not a purely linguistic one, needs to consider semantics seriously. Hans Blumenberg once suggested that the "basic embarrassment of every theology" is "to speak about God constantly without having the right to permit itself to say anything about him" (Blumenberg, 1979, p. 90). Such is the case with innovation. To paraphrase Koselleck (on deeds) once more (Koselleck, 1972): for centuries it was not innovation itself that shocked humanity, but the word describing it.

NOTES

1. A longer version of this chapter is available at www.csiic.ca.
2. Dictionaries of the 17th century emphasized the nontriviality of innovation. For example, Richelet (1680) talks of introducing change "in the State," and Furetière (1690) of introducing change in a thing established "for a long time."

3. A third route was history—of both religion and politics.
4. For the vocabulary used at the time, see the appendix to http://www.csiic.ca/PDF/Old-New.pdf.
5. To Merton, innovation is one of four modes of adaptation of individuals to society. It is a deviant form of behavior, "a departure from institutional norms" (Merton, 1949, p. 144), "the use of institutionally proscribed means" (Merton, 1949, p. 141).
6. "Innovateur" appeared in the sixth edition only (1835): celui qui innove, qui fait des innovations.
7. On 'political revolution', see Arendt, 1963; Koselleck, 1969; Dunn, 1989; Reichardt, 1997. On 'industrial revolution,' see Hardy, 2006. On 'scientific revolution,' see Cohen, 1985.
8. Innovation as 'scientific method' applied to the practical arts is exactly Francis Bacon's vision of science. However, Bacon has never used the term in this sense. To him, innovation had the then-established meaning: introducing change in the established order (religion and politics).
9. The idea of creativity is reserved to God and its "creative power"—although there existed some uses of the latter expression in the 17th century as regards to king's power and man's imaginative faculty.
10. One of the first to put it as such is most probably the French Victor Egger in a thesis on descriptive psychology published in 1881. Egger has distinguished imagination from memory on the basis of innovation. Imagination "innovates" by combining existing elements (coming from memory) in a new way (Egger, 1881, pp. 191–195).
11. Sociologists usually talk of "adoption" (and "diffusion"), whereas economists talk of "introduction" or "implementation."
12. A similar shift as that of the term 'technology,' from a technique to a thing. See Schatzberg (2006).
13. The idea of (and expression) "first introducing" some change or novelty already existed in the 17th century. However, it was not a question of identifying historical origins or originality but a moral one: accusing the person who is responsible for (guilty of) a bad change.

SOURCES (BEFORE THE 20TH CENTURY)

Anonymous. (1789). L'innovation utile, ou la nécessité de détruire les Parlements: Plan proposé au Roi, *La Gazette infernale*. Paris.

Bentham, J. (1787). *Defence of usury; shewing the impolicy of the present legal restaints [sic] on the terms of pecuniary bargains. To which is added, a letter to Adam Smith, Esq. L.L.D. on the discouragement of inventive industry*. Philadelphia: Printed for Mathew Carey, by Lang and Ustick.

Bentham, J. (1793–1795). *A manual of political economy* (Original work published 1843) Retrieved from http://oll.libertyfund.org/index.php?option=com_staticxt&staticfile=show.php%3Ftitle=1922&Itemid=27

Blount, T. (1661). *Glossographia or, A dictionary interpreting all such hard words of whatsoever language now used in our refined English tongue*. London: Tho. Newcombe.

Coles, E. (1677). *An English dictionary explaining the difficult terms that are used in divinity, husbandry, physics, philosophy, law, navigation, mathematicks, and other arts and sciences*. London: Peter Parker.

Defoe, D. (1702). *Essays upon several projects: or, effectual ways for advancing the interest of the nation. Wherein are plainly laid down. The means by which the subjects in general may be eased and enriched; the poor relieved, and trade encreased in the most material branches of it, viz. in constituting seamen to theirs and the nations advantage for encouragement of merchants and merchandizing; for relief of the poor of friendly societies; for discouraging vice, and encouraging vertue; the usefulness; of banks and assurances; to prevent bankrupts; with the surest way to recover bad debts; and many other considerable things, profitable and conducing to the great advantage of the nation in general.* London: printed for Thomas Ballard, at the Rising Sun in Little Britain.

Delepierre, O. (1836). *Aperçu historique et raisonné des découvertes, inventions, innovations et perfectionnements en Belgique, dans les sciences, les arts, l'industrie, etc. depuis les Romains.* Bruges: Félix de Pachtere.

Diderot, D. (1751). Article Enclyclopédie. *Encyclopédie méthodique: logique et métaphysique*, Volume 1 (2): 394–431. Paris : Plomteux.

Dow, C. (1637). *Innovations unjustly charged upon the present church and state. Or An ansvver to the most materiall passages of a libellous pamphlet made by Mr. Henry Burton, and intituled an apologie of an appeale, &c. By Christopher Dow, B.D.* London: Printed by M[iles] F[lesher] for John Clark, and are to be sold at his shop under S. Peters Church in Cornhill.

Egger, V. (1881). *La parole intérieure: essai de psychologie descriptive*, Thèse présentée à la Faculté des lettres de Paris. Paris: Germer Baillière.

Furetière, A. (1690). *Dictionnaire universel.* La Haye: Arnout & Reinier Leers.

Phillips, E. (1658). *The new world of English words, or, a general dictionary containing the interpretations of such hard words as are derived from other languages . . . together with all those terms that relate to the arts and sciences . . . : to which are added the significations of proper names, mythology, and poetical fictions, historical relations, geographical descriptions of most countries and cities of the world. . . .* London: Printed by E. Tyler for Nath. Brooke.

Richelet, P. (1680). *Dictionnaire françois.* Genève: Jean Herman Widerhold.

Sargant, W. L. (1858). *Social innovators and their schemes.* London: Smith, Elder.

Sprat, T. (1667). *The history of the Royal-Society of London for the improving of natural knowledge.* London: Printed by T. R. for J. Martyn, and J. Allestry.

Tarde, G. (1890). *Les lois de l'imitation.* Paris: Seuil. [2001].

Touchard-Lafosse, G. (1822–1824). *Dictionnaire chronologique et raisonné des découvertes, inventions, innovations, perfectionnements, observations nouvelles et importations, en France, dans les sciences, la littérature, les arts, l'agriculture, le commerce et l'industriue, de 1789 à la fin de 1820.* Paris: Louis Colas.

Winslow, H. (1835). *The dangerous tendency to innovations and extremes in education, delivered before the American institute of instruction.* Boston: Tuttle and Weeks.

REFERENCES

Arendt, H. (1963). *On revolution.* New York: Penguin Books.

Barnett, H. G. (1953). *Innovation: The basis of cultural change.* New York: McGraw-Hill.

Berg, M. (1999). New commodities, luxuries and their consumers in eighteenth-century England. In M. Berg & H. Clifford (Eds.), *Consumers and luxury: Consumer culture in Europe, 1650–1850* (pp. 63–85). Manchester: Manchester University Press.

Blumenberg, H. (1979). Prospect for a theory of nonconceptuality. In *Shipwreck with spectator: Paradigm of a metaphor for existence.* Cambridge, MA: MIT Press. [1997].

Carter, C. F., & Williams, B. R. (1957). *Industry and technical progress: Factors governing the speed of application of science*. London: Oxford University Press.
Cohen, I. B. (1985). *Revolution in science*. Cambridge, MA: Harvard University Press.
Daston, L., & Park, K. (1998). *Wonders and the order of nature, 1150–1750*. New York: Zone Books.
Dodgson, M., & Rothwell, R. (Eds.). (1994). *The handbook of industrial innovation*. Cheltenham: Edward Elgar.
Dunn, J. (1989). Revolution. In T. Ball, J. Farr, & R. L. Hanson (Eds.), *Political innovation and conceptual change* (pp. 333–356). Cambridge: Cambridge University Press.
Engell, J. (1981). *The creative imagination: Enlightenment to Romanticism*. Cambridge, MA: Harvard University Press.
Fagerberg, J., Mowery, D. C., & Nelson, R. R. (2005). *The Oxford handbook of innovation*. Oxford: Oxford University Press.
Freeden, M. (1996). *Ideologies and political theory: A conceptual approach*. Oxford: Clarendon Press.
Freeman, C. (1974). *The economics of industrial innovation*. Harmondsworth: Penguin Books.
Gilfillan, S. C. (1935). *The sociology of invention*. Cambridge, MA: MIT Press.
Godin, B. (2010a). *"Innovation studies": The invention of a specialty*. Minerva, Forthcoming.
Godin, B. (2010b). *"Meddle not with them that are given to change": Innovation as evil*. (Working Paper No. 6, Project on the Intellectual History of Innovation). Montreal: INRS.
Godin, B. (2011). *Innovation and politics: The controversy on Republicanism in 17th century England*. (Working Paper No. 10, Project on the Intellectual History of Innovation). Montreal: INRS.
Godin, B. (2012a). How innovation got technological and commercialized innovation. In C. Newfield (Ed.). *Can rich nations still innovate*. Duke University Press. Forthcoming.
Godin, B. (2012b). *The idea of innovation in post-Revolutionary France*. (Project on the Intellectual History of Innovation). Montreal: INRS.
Godin, B. (2012c). *Innovation and culture (Part I): The diffusion controversy and its impact on the study of innovation*, Project on the Intellectual History of Innovation, Montreal: INRS.
Greely, H., Sahakian, B., Harris, J., Kessler, R. C., Gazzaniga, M., Campbell, P., et al. (2008). Towards responsible use of cognitive enhancing drugs by the healthy. *Nature*, 456(7224), 702–705.
Hardy, W. (2006). *The origins of the idea of the Industrial Revolution*. Victoria: Trafford Publishing.
Hunter, M. C. W. (1995). The debate over science. In *Science and the shape of orthodoxy* (pp. 101–119). Woodbridge: Baydell Press.
Kelly, P., & Kranzberg, M. (Eds.). (1974). *Technological innovation: A critical review of current knowledge*. Advanced Technology and Science Studies Group, Georgia Tech, Allanta.
Koselleck, R. (1969). Historical criteria of the modern concept of revolution. In *Futures past: On the semantics of historical time* (pp. 43–71). New York: Columbia University Press [2004].
Koselleck, R. (1996). A response to comments on the Geschichtliche Grundbegriffe. In H. Lehman (Ed.), *The meaning of historical terms and concepts* (pp. 59–70). Washington, DC: German Historical Institute.

Koselleck, R. (2002). The eighteenth century as the beginning of modernity. In *The Practice of Conceptual History* (pp. 154–169). Palo Alto, CA: Stanford University Press.

Koselleck, R. (1972). Begriffsgeschichte and social history. In *Futures past: On the semantics of historical time* (pp. 75–92). New York: Columbia University Press. [2004]

Koselleck, R. (1977). Neuzeit: Remarks on the semantics of modern concepts of movement. In *Futures past: On the semantics of historical time* (pp. 222–254). New York: Columbia University Press. [2004]

Levitt, T. (1966, September). Innovative imitation. *Harvard Business Review*, September, 63–70.

Maclaurin, W. R. (1953). The sequence from invention to innovation and its relation to economic growth. *Quarterly Journal of Economics*, 67(1), 97–111.

Macleod, C. (2007). *Heroes of invention: Technology, liberalism and British identity, 1750–1914*. Cambridge: Cambridge University Press.

Mansfield, E. (1968). *Industrial research and technological innovation*. New York: Norton.

Merton, R. K. (1949). Social structure and anomie. In *Social theory and social structure* (pp. 131–160). Glencoe, IL: Free Press.

Mulgan, G. (2007). *Social innovation: What it is, why it matters and how it can be accelerated*. (Working Paper, Center for Social Entrepreneurship). Oxford: Said Business School.

Norval, A. J. (2000). The things we do with words: Contemporary approaches to the analysis of ideology. *British Journal of Political Science*, 30, 313–346.

Papon, P. (2004). L'invention du Neuf. In *Le temps des ruptures: aux origines culturelles et scientifiques du 21e siècle*. Paris: Fayard.

Pocock, J. G. A. (1975). *The Machiavellian moment: Florentine political thought and the Atlantic Republican tradition*. Princeton, NJ: Princeton University Press.

Radnor, M., Feller, I., & Rogers, E. (Eds.). (1977). *The diffusion of innovations: An assessment*. Evanston, IL: Northwestern University.

Reader, S. M., & Laland, K. N. (2003). *Animal innovation*. Oxford: Oxford University Press.

Reichardt, R. (1997). Révolution. In M. Delon (Ed.), *Dictionnaire européen des Lumières* (pp. 939–943). Paris: Presses universitaires de France.

Richter, M. (1995). *The history of political and social concepts: A critical introduction*. Oxford: Oxford University Press.

Rogers, E. M. (1962). *The diffusion of innovation*. New York: Free Press.

Schatzberg, E. (2006). Technik comes to America: Changing meaning of technology before 1930. *Technology and Culture*, 47, 486–512.

Schmookler, J. (1966). *Invention and economic growth*. Cambridge, MA: Harvard University Press.

Schumpeter, J. A. (1934). *The theory of economic development: An inquiry into profits, capital, credit, interest, and the business cycle*. Cambridge: Harvard University Press.

Schumpeter, J. A. (1939). *Business cycles: A theoretical, historical, and statistical analysis of the capitalist process*. New York: McGraw-Hill.

Schumpeter, J. A. (1942). *Capitalism, socialism and democracy*. New York: Harper.

Shavinina, L. V. (Ed.). (2003). *The international handbook on innovation*. Amsterdam: Elsevier.

Skinner, Q. (2002a). *Visions of politics: Regarding method* (vol. 1). Cambridge: Cambridge University Press.

Skinner, Q. (2002b). *Visions of politics: Hobbes and civil society* (vol. 3). Cambridge: Cambridge University Press.
Smith, G. E., Malinowski, B., Spinden, H. J., & Goldenweiser, A. (1927). *Culture: The diffusion controversy.* New York: Norton.
Stamp, J. (1929). *Some economic factors in modern life.* London: P. S. King.
Stamp, J. (1937). *The Science of Social Adjustment.* London: Macmillan.
Stern, B. J. (1937). Resistance to the adoption of technological innovations. In U.S. National Resources Committee, *Technological Trends and National Policy*, Subcommittee on Technology. Washington, DC: U.S. Government Printing Office.
Stoneman, P. (Ed.). (1995). *Handbook of the economics of innovation and technological change.* Oxford: Blackwell.
Thirsk, J. (1979). *Economic policy and projects.* Oxford: Oxford University Press.
Thorndike, L. (1957). Newness and novelty in 17th-century science and medicine. In P. P. Wiener & A. Naland (Eds.), *Roots of scientific thought: A cultural perspective* (pp. 443–457). New York: Basic Books.
Thorpe, C. D. (1937). Addison and some of his predecessors on 'novelty'. *PMLA*, 52(4), 1114–1129.
U.S. Department of Commerce. (1967). *Technological innovation: Its environment and management.* Washington, DC: U.S. Government Printing Office.
Usher, A. P. (1929). *A history of mechanical inventions.* New York: Dover.

4 The Unintended and Undesirable Consequences

Neglected by Innovation Research

Karl-Erik Sveiby, Pernilla Gripenberg, and Beata Segercrantz

In his review of innovation literature almost 30 years ago Rogers (1983) noticed that only 0.2% of innovation diffusion articles addressed *consequences* of innovation. Still today, the study of consequences seems to be marginalized across all main areas of the rapidly emerging field of innovation research. The purpose of this chapter is to study research on unintended undesirable consequences of innovation. Our research questions are: How much of innovation research has captured undesirable consequences[1] of innovations? What subjects have been covered in this research? What kinds of consequences have been found? Who have been affected by these consequences? What methodologies have been used to study consequences of innovation?

To answer our research questions, we conducted two searches of articles[2] in peer-reviewed journals and one screening of the six most influential handbooks in the field. We found that only 0.1% and 0.4% articles respectively and no handbook chapters were devoted to unintended and undesirable consequences of innovation. Based on these reviews, our conclusion is that innovation research is mainly concerned with what is directly measurable as economically beneficial for innovating firms and change agents, whereas nonbeneficial consequences for other stakeholders in society are disregarded (with one exception: unemployment).

The chapter is organized as follows: First, we develop a theoretical framework for studying unintended consequences of innovation. Then, we present the results of the first literature review on what undesirable consequences of innovation have been found and position them tentatively in relation to Fagerberg and Verspagen's (2009) mapping of 'innovation studies.' We then validate our first search results by conducting a second article search,[3] this time focused on the five most influential journals in 'innovation studies,' and reporting the reading of the six most influential innovation handbooks. Finally, we discuss the findings and draw some conclusions concerning research in the scientific field of innovation.

UNINTENDED CONSEQUENCES AND INNOVATION

Contributions from four theories have inspired the theoretical framework of this study. These theories are economic theory, communication theory (diffusion of innovation), Merton's sociological theory, and stakeholder theory. Unintended consequences have been a fundamental element in economic theory since Adam Smith's work in the 18th century. According to Smith, prosperity for society as a whole is generated by individuals, who are led 'by an invisible hand to promote an end which was no part of his intention' (Smith, 1789).

The economist Joseph Schumpeter (1939) sees 'innovation, together with all their effects, and the response to them by the economic system' as the primary generator of growth in the economy. Innovation is 'the outstanding fact in the economic history of capitalist society . . . and it is largely responsible for what we at first sight attribute to other factors' such as industrial processes (1939, p. 86). In the Schumpeterian economic theory, the effects of innovations are 'disruptions' in the economy, which become the main cause of both the growth of new industries and companies and the demise of old ones. For some, the direct and indirect consequences will be desirable, others will suffer; such cycles are part and parcel of the capitalist economic system. Schumpeter's theory has been widely criticized, and the role of innovation as the primary generator of growth has been reduced somewhat in later economic theory. Yet his definition of innovation as 'the setting up of a new production function' (p. 87), in which he distinguishes between product and process innovation still forms the basis of later definitions. F. A. Hayek, for example, argues that the unintended consequences of individuals' actions create the 'spontaneous order,' which is what characterizes the capitalist market economy (see Barry, 1982).

Studies on general effects and consequences of innovation are conducted within at least two theoretical lenses. The first one is the literature streams discussing effects of technological change within the fields of technology assessment (TA) and the social shaping of technology (SST). Those literature streams study technological change from processual perspectives and are discussed by Hasu and colleagues in chapter 5 of this book. They conclude, however, that so far TA and SST processes have not been used much in literature to address undesirable effects of innovations.

The second stream studies diffusion of innovation, originally an anthropological concept (Godin, 2011), but later incorporated more generally in sociology and, in the 1960s, also in economics. In economics, diffusion theory is regarded as the third stage in the Schumpeterian innovation 'trilogy' (after invention and product development), when the new products and processes spread across the potential market and various impacts emerge and become known (Stoneman, 1995, pp. 2–4). The innovator's representatives are regarded as driving forces with the intention to spread or increase the usage of the innovation among adopters. Hence, economic theory defines diffusion from the innovator's perspective. This theory primarily focuses on the economic impacts on firms and industry, and unintended and/or undesirable consequences are not among the main themes (Lissoni & Metcalfe, 1994).

Sociological approaches to innovation, on the other hand, typically define diffusion from the adopters' perspective and sees diffusion as 'the process by which an innovation is communicated through certain channels over time among the members of a social system' (Rogers, 1983, pp. 1–7; see also chapter 2 of this book). Sociological diffusion theory recognizes unintended and indirect consequences on groups of individuals and is commonly used to evaluate social effects of marketing methods (Lowrey, 1991; Rogers, 1976) and to analyze introduction of new processes in organizations, (Feller & Menzel, 1977; Johns, 1993; Kitchener, 2002), in particular, adoption of new ICT technology (Irwin, Hoffman, & Geiger, 1998; Robertson, Swan, & Newell, 1996). Rogers's version of diffusion theory applies a taxonomy consisting of three dichotomies of consequences of innovation: desirable versus undesirable, direct versus indirect, and anticipated versus unanticipated. The taxonomy is applied from the perspective of the 'members of the system,' that is, both change agents and adopters.

Today Rogers work is marginalized in the innovation field as both we and Fougère and Harding (see chapter 2 of this volume) conclude, and we argue that many issues raised by Rogers are in great need of more research today. For the purpose of this chapter, we therefore draw on the taxonomy developed by Rogers, although it has not contributed much in explaining why actors are rarely able to anticipate indirect consequences. We suggest that one way forward is to amend diffusion theory with the sociologist Robert Merton's (1936) concept of limiting factors and with stakeholder theory.

The first systematic analysis of the concept of unintended consequences can be found in the paper "Unanticipated Consequences of Purposive Action"[4] by Merton (1936). Although 'unanticipated consequences' and 'unintended consequences' are not synonyms, later authors use the two terms interchangeably (see, for example, Hill, 2003; Rogers, 1983). Merton also, in later a work (Merton, 1968, pp. 120, 182), treats the term anticipated as the antonym of unintended. However, following Saha (1998) and Baert (1991), we distinguish between the two terms because an unintended consequence can be anticipated. For instance, if I intentionally drive through a red light, I can anticipate other drivers' angry honking, even if it was not my primary intention with my action.

Merton defines 'consequences of purposive action' as both those that are exclusively the outcome of the action, we call them here direct consequences, and those that are the consequences of 'the interplay between the action and the objective situation,' that is, causally related, indirect consequences according to the typology of Rogers (1983). Merton distinguishes five factors that limit an actor's possibility to anticipate both direct and indirect consequences. The first is general lack of foreknowledge; it is a fact of life that humans often (have to) act without full knowledge about consequences that (may) follow.

The second limiting factor is actors' assessment errors in their assumptions, in their selection of a course of action, or in their execution of acts. Merton (1936, 901) mentions habit as a source of error: The common fallacy to believe 'that actions, which have in the past led to the desired outcome, will continue to do so.'

Merton labels the third factor the 'impervious immediacy of interest.' By this he refers to actors desiring the beneficial consequences of an action so much that they are adamant and blind to any other temporal or spatial consequences. We choose the term *myopia* for this factor, as suggested by McAulay (2007).

We call Merton's fourth source of unintended consequences *ideology*, which is described as follows; 'there is no consideration of further consequences because of the felt necessity of certain action enjoined by certain *fundamental values* [our emphasis]' (Merton 1936, 903). His example is the Protestant ethic of hard work and asceticism, which 'paradoxically leads to its own decline through the accumulation of wealth and possessions.'

The fifth factor forms a feedback loop as Merton explains: When predictions of future social developments are made public they become 'a new element in the concrete situation,' thereby changing the course of history. Merton developed this notion further in later works (Merton, 1968) and coined the often-cited phrase 'self-fulfilling prophecy' with its opposite, the self-defeating prophecy, which is the term we use here. For instance, because organizational change initiatives have failed in the past, following change initiatives are met with cynicism by employees, thereby further increasing the risk of failure.

Rogers does not cite Merton's theories, but Merton's concept of unintended consequences has been applied frequently by social scientists and political economy scholars outside the innovation field to study and criticize indirect, unintended effects of social policy (Andersen & Serritzlew, 2007; Hill, 2003; Legge, Jr., 1983), economic policy (Glinavos, 2008; Holzer & Millo, 2005), and new legislation (Glinavos, 2008; Iyengar, 2008; Rose, Clear, & Ryder, 2001). In marketing, critique of the unintended consequences of advertising have a long history (Calfee, 1987; Fry & Polonsky, 2004). Also, introduction of new information systems (Binbasioglu & Winston, 2004; Harrison, Koppel, & Bar-Lev, 2007) and effects of organizational change (Fairhurst, Cooren, & Cahill, 2002; McNamara, Moon, & Bromiley, 2002; White & Ramsey, 1978) have been studied. The concept has by now reached a taken-for-granted level in the social sciences; it is frequently used, but rarely discussed theoretically.

Summary of Theoretical Framework

The three theories list a number of 'groups or individuals who can affect or are affected' by the introduction of an innovation, thereby fulfilling the

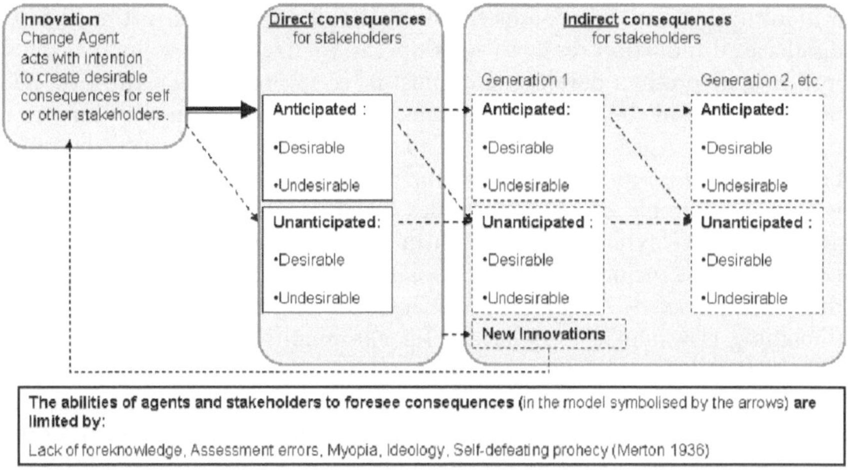

Figure 4.1 Intended and unintended consequences of innovation.
Source: Adapted from Rogers (1983), Merton (1936), and Freeman (1984).

definition of stakeholders (Freeman, 1984, p. 46). A far from comprehensive list contains at least the following stakeholders: entrepreneurs, inventors, competitors, businesspersons (managers), shareholders, employees, bankers (Schumpeter, 1939), change agents, sponsors, adopters, consumers (Rogers, 1983), actors, other people or bystanders, social structure, and civilization (Merton, 1936). Certain stakeholders tend to more frequently affect the introduction of innovations, whereas others are more often affected by them. As with most classifications, the two roles overlap to some degree; adopters may apply an innovation in unanticipated ways, thereby affecting it. Also, desirable and undesirable effects overlap and coincide; some adopters will find the innovation desirable, features that some consumers love others may hate. Our study focuses primarily on those negatively affected by an innovation; those who find its consequences undesirable.

Our analytical framework can now be summarized in Figure 4.1. Diffusion of innovations is influenced by Merton's limiting factors. They limit both change agents' and adopters' abilities to anticipate direct and indirect consequences of their actions for other stakeholders. Our research interest in this chapter is in the undesirable consequences, both direct and indirect.

WHAT UNINTENDED UNDESIRABLE CONSEQUENCES OF INNOVATION HAVE BEEN FOUND?

To find out what unintended and undesirable consequences of innovation have been found in previous research and how the subject has been studied, a literature review was conducted using Boolean searches in the EBSCO database. 'Innovation research' was operationalized so that 'innovati*' or 'NPD' (new product development) must be so important that the term had been included in the title of an article.[5] A search based on that definition conducted on April 23, 2009, yielded 26,300 articles. Because the number of articles on unintended and undesirable consequences of innovation within this sample of 26,300 articles became even smaller than we had anticipated, we extended the search in EBSCO further by including more terms and also included Google Scholar. The combined searches yielded a first selection of 477 articles that mentioned consequences of innovation (including synonyms). A more careful examination of the abstracts was done through several readings by different members of the research team to establish whether the articles fulfilled the criteria of studying consequences of innovation. This examination reduced the number of articles to a total of 56. A final selection of 26 articles on unintended and undesirable consequences of innovation was then selected for further review and analysis. See table 4.1 for the search terms.

The 26 articles were analyzed in terms of their subject area and type of innovation, their research methodology, the undesirable consequences of the innovation, and the change agents and stakeholders. We also categorized, as far as possible, the factors that limit a change agent's anticipation of the consequences of an innovation. The results of the analysis can be seen in table 4.3.

Table 4.1 Search Terms Applied and Number of Articles

Search terms	**Boolean search phrases**	**Totals**	**Selected based on Abstracts**	Final selection
Innovati or NPD	TI innovati* or TI NPD in Title	26 300		= 100%
First EBSCO searches	(TI innovati* or TI NPD) and combinations of indirect, negative, consequences, results and effects in Abstract	255	27	12 (0.05%)
Extended EBSCO + Google Scholar full text searches	Relevant combinations of indirect, negative, unexpected, unintended, unanticipated, effects, side effects, results, impact*, consequences. Full text	222	29	14
Total articles analyzed	(Some overlap between Google Scholar and EBSCO)	477	56	26 (0.10%)

Article Review Findings

Among the innovations covered by the 26 reviewed articles, information technology innovation as organizational adoption of new ICT (ICT or information systems (IS) processes) is the most common (in 7 articles). Other distinct areas are technical innovations, innovations in food production and agriculture, manufacturing processes, banking products, institutional and managerial innovations, human resource management, and business model innovation. Methodologically, the articles comprise a vast variety of empirical methods, such as case studies, surveys, and econometric modeling of panel data. Analyses of time series are common. Case analyses are mainly comparative, often with a longitudinal element to cover both changes and their consequences. Survey data are often combined with other data, such as panel or archival data, or conducted in two steps. Many articles combine several methods and data sets. One fifth of the articles are theoretical, based on extensive literature reviews.

Perhaps unsurprisingly, society, social structure, and bystanders are affected by the *indirect* undesirable effects of innovations. *Direct* undesirable consequences of innovations affect mainly organizational employees and consumers. Managers are the most numerous (in 10 articles) change agents affecting innovation-generating undesirable consequences, followed by entrepreneurs and inventors (5). Other change agents are investors, shareholders, and academics. Myopia is most commonly featured as the factor that limits actors' ability to foresee undesirable consequences. Introduction of new organizational processes are often hindered by managerial myopia. Managerial myopia may even cause reduced innovation in an industry or increased reliance on development aid for society.

The articles in our sample give several examples of assessment errors due to change agents' pro-innovation bias. For example, in banking, low-level innovative financial products can be more profitable than highly innovative ones; in franchising, imitation can be more efficient than innovation; and introductions of ICT often fail due to 'mindless' implementation. Smoller, Osborn, and Price (2001) gave a prescient warning for a financial crisis caused by innovation already in 2001, whereas Nicholas (2008) links both the 1990s and the 1920s stock market crashes to the arrival of new technologies. Change agents' tendency to believe that 'the only way to reduce negative consequences of innovation is to continue innovating' (Pinto, 1996) can be the cause of assessment errors. For instance, changes in behavior and lifestyle can be more effective in reducing CO_2 emissions (caused by older innovations) than technical innovation, as one author in our sample argues.

We also compared our sample with the innovation studies field, as described by Fagerberg and Verspagen (2009), in terms of geographical spread of authors, type of journals, and main sources of inspiration. Our sample authors are mainly U.S. based (in 14 articles). Other nationalities represented are the U.K. in 4 articles; other European nationalities in 3 articles (France, Germany, and Greece); cross-national collaborations in 4 articles (U.K. and U.S., U.K. and Italy, U.S. and Singapore, U.S. and South Africa); and Australia in 1 article. Hence, compared to the majority of the mainstream innovation scholars who, according to Fagerberg and Verspagen (2009), are European based; our sample seems to be different. The sample also seems to differ from the groups of innovation researchers identified. The majority of the articles are published in management and business journals (including also innovation management and strategic management) with 13 articles. Eight articles are published in economic journals (including also agricultural, developmental, and labor economics). Three articles are found in the IS/ICT literature, and two articles are found in marketing journals.

In the sample, the main source of inspiration in terms of citations is Rogers's (1962; 1983; 1995; 2003) *Diffusion of Innovations*, with 13 citations in 9 of the 26 sample articles. Our analysis of the citations and main sources of inspiration to the sample article authors can be seen in table 4.2. In the analysis, we analyzed all references in the 26 reference lists. We analyzed both how many times a central author had been cited, as well as in how many of the sample articles a central author's work had been cited. The authors in the table are sorted by the column 'Citing articles in sample,' which shows how many of the 26 sample articles had referred to that particular central author, indicating the spread of the influence in the sample articles.

In sum, our sample could perhaps be called a subbranch of Fagerberg and Verspagen's (2009) management cluster. Their article identified five clusters in innovation studies of which the management cluster is one. It is mainly active in North America, and it is using Rogers as the most important source of inspiration, rather than Freeman (1974) or Schumpeter constituting the most dominant clusters. Also DiMaggio and Powell's work on institutional isomorphism emerge as important to our sample authors, whereas their work is not at all recognized by Fagerberg and Verspagen as a source of inspiration for the majority of innovation researchers. Only one of the articles we found can be said to apply something similar to the TA and SST literature stream. The articles that study undesirable consequences of innovation are not found in the most influential innovation research journals, except for two articles published in *Research Policy*; the most important journal for 45.6% of innovation scholars in Fagerberg and Verspagen's study.

Do 'Innovation Studies' Consider Undesirable Consequences?

Because Fagerberg and Verspagen's (2009) article was published too late to guide our first study, we decided to use their mapping of 'innovation studies' for validation instead. We did an article search on the five journals[6] ranked as the most important by innovation researchers and a reading of

Table 4.2 Citations and Main Sources of Inspiration

Author	Citing articles in sample	Main or most recent work cited	Times cited	Total # citations in sample articles
Rogers, E. M.	9 (35%)	(1962, 1983, 1995, 2003). *Diffusion of innovations.* New York: Free Press.	12	14
DiMaggio, P. J. & Powell, W. W.	6 (23%)	(1983). "The Iron Cage Revisited: Institutional Isomorphism and Collective Rationality in Organizational Fields," *American Sociological Review* (48), pp. 147–160.	4	7
Cohen, W. & Levinthal, D.A.	6 (23%)	(1990). "Absorptive Capacity: A New Perspective on Learning and Innovation," *Administrative Science Quarterly* (35), pp. 128–152.	4	7
Nelson, R. & Winter, S.G.	5 (19%)	(1982). *An Evolutionary Theory of Economic Change,* Cambridge, MA: Harvard University Press.	4	12
Rosenberg, N. & Mowery, D.C.	5	(1998). *Paths of Innovation: Technological Change in Twentieth-Century America.* Cambridge: Cambridge University Press.	2	8
Schumpeter, J. A.	5	(1934). *The Theory of Economic Development.* Cambridge, MA: Harvard University Press.	3	6
Teece, D. J., Pisano, G., & Shuen, A.	5	(1997). "Dynamic capabilities and strategic management," *Strategic Management Journal* (18), pp. 509–533.	2	6
Scherer, F. M.	5	(1980, 1990). *Industrial Market Structure and Economic Performance.* Chicago, IL: Rand McNally.	5	5

Continued

Table 4.2 Continued

Porter, M. E.	4 (15%)	(1985). *Competitive Advantage: Creating and Sustaining Superior Performance*. New York: Free Press.	1	8
Kogut, B. & Zander, U.	4	(1992). "Knowledge of the Firm, Combinative Capabilities, and the Replication of Technology," *Organization Science* (3), pp. 383–397.	2	5
Tornatzky, L. G., Fleisher, M., & Chakrabarti, A. K.	4	(1990). *The processes of technological innovation.* Lexington, MA.: Lexington Books.	2	5
Cooper, R.	4	(1990). "Information technology implementation research: A technological diffusion approach," *Management Science* (36)2, pp. 123–139.	3	4
Chandler, A. D.	4	(2002). The opportunities for business history at the beginning of the 21st century. In F. Amatori & G. Jones (Eds.), *Business history around the world.* Cambridge: Cambridge University Press.	1	4

the six handbooks (see table 4.5.) chosen by Fagerberg and Verspagen to identify experts in the field (Fagerberg, 2010). The journal search yielded five articles, or 0.4% of the population, which is four times more than our first search's 0.1%, but still an immaterial proportion of the population. The handbook reading did not yield a single chapter. Both the second journal study and the handbook reading therefore validate the result of our original study: Undesirable consequences of innovation are neglected by innovation research. There is one encouraging new observation, nevertheless. Three of the five articles found in the second journal search are very recent, from 2008, and all but one are newer than the handbooks. Are they a 'weak signal' of changing attitudes? See details of the analysis that follow.

Table 4.3 Articles on Undesirable Consequences of Innovation

Citation	Subject area/ Type of Innovation	Main contribution	Research methodology	Undesirable consequences			Limiting factors
				Type of consequences	Type of agents	Type of affected stakeholders	
Abrahamsson (1991)	Innovation in general	Ineffective innovations persist due to fads and fashions	Theory development	Indirect: Lower productivity		Society	
Alam (2003)	Financial products	Moderately innovative products more successful than highly and low innovative	Two-step survey	Direct: Innovation for innovation's sake, not profitable	Managers in NPD process	Shareholder	Assessment error, fundamental values
Aubert, Caroli, & Roger (2006)	IS/ICT processes	Innovative firms tend to be biased against older works	Econometric modeling on national level time series data	Direct: Wage loss and loss of employment opportunities for older workers (50+)	Managers	Older employees at the workforce level	Ignorance
Calantone, Chan, & Cui (2006)	NPD in U.S. chemical, biotechnical, and pharmaceutical industries	Level of innovativeness has no direct effect on new product profitability	Survey structural equation modeling	Direct: Need to adapt to customer Indirect: no profit for shareholder	Change agent (NPD, PD, line and product managers)	Consumers, shareholders	Myopia

Continued

Table 4.3 Continued

Citation	Subject area/ Type of Innovation	Main contribution	Research methodology	Undesirable consequences			Limiting factors
				Type of consequences	Type of agents	Type of affected stakeholders	
Chigona & Licker (2008)	IS/ICT processes	Poor area development: adoption of communal computing facilities in poor area libraries	Multiple method data collection; Cases compared and abductively tested	Direct: Decrease in library loans, increase in noise level Indirect: Skills divide	Managers, scientists	Employees, social structure, bystanders	Ignorance
Clark, Bennett, Burcher, Newell, Swan, & Sharifi (1992/1993)	IS/ICT processes	Why IS/ICT implementations fail; Users should be involved more closely in the initial phase of initiation	Case study on computer-aided production management introduction	Direct: Inefficiency, frustration in organizations	Change agents	Adopters, employees	Self-defeating purpose, myopia
Desmarchelier & Szabo (2008)	Food production	Overview of innovations that raise health concerns	Multiple cases as examples	Direct: Wide range of health consequences	All in food production value chain	Employees, consumers, society	Ignorance, assessment errors, myopia
Evangelista & Savona (2002; 2003)*	IS/ICT processes	IS/ICT innovation in large service firms decreases employment in overall Italian service sector	Econometric modeling on Italian time series data	Indirect: Increased unemployment	Managers	Employees, social structure	Fundamental values, myopia

*other works than main works included

Farell & Runge (1983)	Agriculture	New Deal institutional innovations in 1930s turned U.S. farms into big, highly leveraged, unwieldy administrative structures	Historical case as example	Indirect: Small farms disappear	Government	Farmers, society	Ignorance, ideology
Fichman (2004)	IS/ICT processes	How to improve IS/ICT adoption research and avoid nonbeneficial ICT implementation	Literature review	Direct and indirect: How to avoid negative effects in general		Adopter organization	Assessment errors, myopia
Goeschl & Swanson (2000)	Food production/ agriculture	IP protection of GURT (genetic use restriction technologies) seeds reduces diffusion in development countries	Time series econometrics from 1961–1999	Indirect: Slower agricultural productivity growth, more development aid required	Managers	Small farmers in developing countries	Myopia, self-defeating purpose, ideology
Gross (2005)	General technological change	Introduces Simmel's sociology of science; New technology always has both expected and unexpected consequences	Theory development			Society	
Herring & Roy (2007)	Energy consumption	Rebound effects cancel effects of innovations to reduce energy consumption; Only changes in behavior and lifestyle work	Metastudy on historical cases	Indirect: on climate change	Inventors, scientists	Civilization	Assessment errors

Continued

Table 4.3 Continued

Citation	Subject area/ Type of Innovation	Main contribution	Research methodology	Undesirable consequences			Limiting factors
				Type of consequences	Type of agents	Type of affected stakeholders	
Hull & Kaghan (2000)	Innovation in general	Argues need for more critical studies of innovation; Suggests e.g. critical management studies and science and technology studies	Editorial/ introduction to special issue				
Koeller (1995)	Manufacturing industries	Large firms' innovation increases industry concentration	Regression on time series data	Indirect: reduces small and medium sized enterprises (SME) innovation	Entrepreneurs, managers	Employees, competitors	Self-defeating purpose
Koeller (2001)	Manufacturing industries	Innovation in manufacturing industries and union membership	Econometric model, time series data	Indirect: Union membership	Managers	Employees, social structure	Myopia
Kossek (1989)	Human resource process	Major differences among employees in acceptance of human resource (HR) innovations	Diagnostic method, regression on survey data	Direct: Increased cynicism	Managers	Employees	Myopia, self-defeating purpose
Nicholas (2008)	Innovation in general	Innovation caused stock the market run-up, followed by crash in 1930s	Regressions on time series, panel data	Direct: Stock market boom Indirect: Leading to recession	Investors, share-holders	Investors, share-holders	Myopia, assessment errors
Ramdani & Kawalek (2007)	IS/ICT processes	Review of ICT adoption models	Theory development				

Schroeder (1990)	Manufacturing	One innovation's impact on firm competitiveness; Develops theory	Longitudinal study on 5 cases over 25 years	Both positive and negative Schumpeterian effects	Entrepreneurs, managers	Entrepreneurs, competitors	
Scranton (2006)	Innovation in general	Emphasizes uncertainties inherent in innovation	Theory development, Historical				
Skuras, Tsegenidi, & Tsekouras (2008)	Manufacturing	Innovation competes with other risky investments in SMEs	Survey and statistical analysis of official financial records	Direct: Innovation reduces SMEs' probability to invest in capital assets	Entrepreneur	Society	Assessment error
Smith & Skalnik (2003)	Food production	Growth hormone introduction to increase milk production benefits change agent only	Comparative case study of U.S. vs. E.U. stands against bovine growth hormone in milking cows	Indirect: Unknown long-term risks of biotechnology	Large corporations (e.g., Monsanto)	Consumers, social structure, farmers	Myopia
Smoller, Osborn, & Price (2001)	Financial industry	Risks in new derivatives are underestimated	Case analysis. Combines tech. change and behavioural economics theory	Indirect: Collapse of the financial system	Bankers	Shareholders, society	Myopia, Ignorance, Assessment error
Swanson & Ramiller (2004)	IS/ICT Processes	How organizations can be mindful rather than mindless in adoption of ICT	Theory development Literature review	How to avoid negative effects of innovation	Managers	Adopter organization	
Szulanski & Jensen (2008)	Franchising	Compares replication of franchise concept or adapting it (=innovation)	Multimethod: Archival data and survey to franchise networks	Direct: Innovating from scratch reduces franchisee growth	Entrepreneur (franchisees)	Change agent, entrepreneur	Ignorance

Table 4.4 The Proportion of Articles on Undesirable Consequences in the Five Top Journals of 'Innovation Studies'

Boolean search phrases	# of articles	Proportion
The top five journals in 'Innovation studies' all articles in EBSCO database	3,909	= 100%
AND (innovation OR NPD OR technolo* change* OR organizational change OR corporate change OR structural change) in Abstract	1,300	
AND [Relevant combinations of]: indirect, adverse, negative, unexpected, harmful, pollut*, undesirable, unwanted, dangerous, destructive, environmental, unintended, unanticipated, effect*, side effect*, result*, impact*, consequence*, output*, externalit*, pollutant* in Abstract	31	
Thereof articles studying the previous effects caused by innovation	5	= 0.4%

THE FIVE MOST IMPORTANT JOURNALS

The first journal search in 2009 and suggestions from colleagues had yielded more synonyms of 'consequences,' 'undesirable' and 'innovation' and the second search was hence amended accordingly, see table 4.4. The five journals contained a total of 3,909 articles at the time of search on August 18, 2011. The expanded search term list was applied on abstracts with the intention to catch more articles than the first search while maintaining the requirement that the topic must be of importance according to the author. The search yielded a gross list of 31 articles.

Only five articles fulfill the criteria, that is, only 0.4% of the articles in the five top journals in 'innovation studies' study anything negative or undesirable in relation to impacts of innovation on society. Three of them had been identified already in our first sample of 26 in table 4.3 and two were added: Winkler (2005), who discusses how to include potential harmful outputs in economic models, and Giuri, Torrisi, and Zinovyeva (2008), who study effects of organizational change on employment in Italian manufacturing firms. Innovation is the dependent variable in 20 of the 31 articles, that is, they study effects that have a negative impact on innovation, rather than negative effects that follow innovation.

SIX HANDBOOKS REPRESENTING THE EXPERTISE

The screening of the 181 chapters in the 6 innovation handbooks did not identify a single chapter devoted to undesirable consequences of innovation (see table 4.5). Four of the six handbooks apply economic theory and one

(Tidd, Bessant, & Pavitt) takes a managerial perspective. The unit of analysis is generally the firm, and the authors discuss how to improve innovation and its impact on firm performance or society's (economic) 'welfare.'

The handbooks are not entirely void of alternative perspectives, however. The *International Handbook of Innovation* (2003) is much broader than

Table 4.5 Coverage of Undesirable Consequences* in Innovation Handbooks

Authors/ (Perspective)	**Title**	**Chapters**	**Comment**
Dosi, G., et al. (1988). (Economics)	*Technical change and economic theory*	27	**Lundwall** discusses national systems for innovation; one example of a Danish 'unsatisfactory innovation.'
Stoneman, P. (1995). (Economics)	*Handbook of the economics of innovation and technological change*	13	**Beath et al.** mentions negative externalities on early adopting firms. **Petit** writes a chapter on theories of employment effects of innovation.
Dodgson, M., & Rothwell, R. (Eds.). (1996). (Economics)	*The handbook of industrial innovation*	35	**Skea's** chapter discusses effects on firms due to consumers' and governments' environmental concerns. **Lissoni & Metcalfe's** overview of main themes in diffusion theory does not include undesirable consequences.
Shavinina, L. V. (ed.). (2003). (None)	*International handbook on innovation*	71	**Tudor** mentions in a final conclusion chapter that critical perspectives on innovation may become more common.
Fagerberg, J., Mowery, D. C., & Nelson, R. R. (2006). (Economics)	*The Oxford handbook of innovation*	22	**Pianta's** chapter highlights both positive and negative employment effects. **Miles** discusses negative effects of regulation on innovation.
Tidd, J., Bessant, J., & Pavitt, K. (2005). (Management)	*Managing innovation: Integrating technological, market and organizational change*	13	One page on "innovation for sustainability."
		181	No chapter is entirely devoted to undesirable effects/ consequences.

*Undesirable consequences are defined as a collective of all search terms—same as in Table 4 .4

the other handbooks with contributions also from psychology, but it still fails to include a chapter with a critical perspective or one on undesirable consequences among its 71 chapters. The other exception is two chapters, which discuss employment effects caused by technological change. Petit (in Stoneman, 1995) makes an overview from the Middle Ages to today of the debate and illustrates with both amusing and disturbing examples the failure of economic modeling to predict employment effects of innovation. Ten years later, Pianta examines the literature (in Fagerberg et al., 2006) and concludes that: 'There is no automatic mechanism ensuring that national or regional economies fully compensate for innovation-related job losses.' The key issue is instead 'the rate at which technological innovation and diffusion eliminate jobs, versus the pace at which new economic activities create new jobs.' Two further minor contributions are worth highlighting: Lundwall mentions (in Dosi et al., 1988) an 'unsatisfactory innovation' as a cautionary example for regulators in a discussion on national systems for innovation, and Beath and colleagues briefly mention negative externalities in terms of the losses a firm may incur by being a too early adopter (in Stoneman, 1995).

Environmental issues, harmful products or issues for society outside the industry, or firm contexts are rarely raised as topics, and when they are, innovation is the dependent variable. Two examples: Consumer concerns, NGOs, and environmental regulations are mentioned in terms of obstacles for innovating firms (Miles, in Fagerberg et al., 2006), or as opportunities for innovation (Tidd et al., 2005). The fact that the issues and concerns may be caused by innovation in the first place is not a topic.

DISCUSSION AND CONCLUSIONS

In his overview of innovation research in the *Oxford Handbook of Innovation* (2006, p. 20) Fagerberg lists three 'desirable consequences' of innovation that make policy makers and business leaders interested in promoting it. First, innovation is crucial for long-term economic growth; second, innovation tends to cluster in certain more rapidly growing industries/sectors; and, finally, innovation is a powerful explanatory factor behind differences in performance between firms, regions, and countries. As Fagerberg argues (p. 20): 'Firms that succeed in innovation prosper, at the expense of their less able competitors.' However, he continues: 'in spite of the large amount of research in this area during the past fifty years, we know much less about why and how innovation occurs than what it leads to.' Presumably the 'we' refers to researchers in the field of 'innovation studies.'

We beg to differ. Our view is that innovation research, defined more broadly (as in our first article search) or more narrowly (as 'innovation studies'), has a very poor understanding of at least one aspect of what innovation leads to: unintended undesirable consequences for other stakeholders than the innovating firms and the change agents. This is because they are given practically no attention. We base our conclusion on two literature searches and one reading of the six most influential handbooks in the field of innovation studies. We found only 0.1% to 0.4% of articles devoted to the topic and no innovation handbook chapters. Admittedly, the validity of our two article searches is limited; they do not cover the whole universe of innovation research in all databases and not all conceivable search terms. Other search terms used at other times in other databases will yield different selections of articles and the exact proportion will vary. Given the extremely low proportions that we have found it is, however, highly unlikely that other searches will yield proportions high enough to alter the conclusion.

We have studied a considerable amount of literature in this process and we suggest two main reasons for the lack of attention. One is that innovation research seems to be built on a fundamental—generally implicit—assumption that 'innovation is good.' Whether the assumption is characterized as ideology or myopia, the outcome is the same; the assumption limits the ability and interest of researchers as well as decision makers and change agents to consider unintended undesirable consequences of innovation. The strength of the bias is underlined by our finding that the amount of research on unanticipated, undesirable consequences of innovation in the field of innovation studies is still as marginal as it was in the 1960s when Rogers made his first review of innovation diffusion research.

The other reason for lack of research on undesirable consequences in innovation research appears to be a separation of discourses of desirable and undesirable consequences. The articles we have found are an eclectic selection of undesirable consequences of innovation from a variety of industries. Just as important, therefore, are research fields and discourses, which are not represented. There are, for instance, no articles about environmental consequences, none on side effects of medicines or failed product introductions, only one about undesirable consequences of new legislation (U.S. government), and only two about undesirable consequences for consumers. Only one article applied TA/SST frameworks. Undesirable consequences of innovation exist with all types of innovations; the few studies we have found prove the point. Innovation studies, however, seems to have implicitly isolated itself from considering them. There is one notable exception: We have found a few studies on the effects of innovation on employment among the articles, and they show up also in the handbooks.

The fact that innovation studies do discuss negative employment effects sheds some light on the nature of their isolation. It may have to do with the dominance of economic theory in the field; labor has been an important input factor in economic modeling since the 1800s. Because effects and consequences of innovation for consumers, individual health, society, ecology, and the environment have never been introduced as part of economic theory, they are, hence, still implicitly excluded from the research agenda. These have instead become topics in many other discourses and are constructed with other terminologies and theories and from other disciplines than economics. These other discourses do indeed study the problems caused by innovation, but the scholars do not see themselves as producing 'innovation research,' and even if they do, their articles are rarely accepted by the economics-dominated top journals in 'innovation studies,' or else their articles would have shown up more prominently in our article searches.

The separation can therefore be seen as an indication of paradigmatic isolation. A further indication of this is our observation that innovation is treated as the dependent variable in most articles we have found. Anderson, De Dreu, and Nijstad (2004) made the same observation, and we agree with them that innovation research appears to have become 'routinized.' Hence, we suggest that innovation studies would benefit creatively from stepping out of its self-imposed economic boundaries.

Another implication is that the vital connection between innovation and its undesirable consequences 'gets lost' also in policy making. The separation of research discourses tends to be mirrored by compartmentalization on policy level. To stylize: the ministers of industry and economy are guided by innovation studies and hence become 'ministers for desirable consequences of innovation,' whereas the heads of the health, social, and environment departments can call themselves 'ministers for the undesirable consequences of innovation.'

Another implication is a case of self-defeating prophecy. As policy makers and business leaders receive little practical guidance from innovation researchers on how to consider undesirable consequences, they continue to add more of them, thereby never reaching the full net value of innovation for society.

A further potentially dangerous implication is that the lack of research on undesirable consequences in 'innovation studies' allows the concept to retain its positive buzz among policy makers, because it is rarely 'blamed' for anything negative or undesirable—not even unemployment these days. This creates another self-defeating prophecy: Because the connection between innovation and undesirable consequences is not addressed by innovation research, funding bodies see little point in funding such studies, thereby further reducing them.

What can we do to remedy the situation? This chapter is an effort to raise awareness of the issue among innovation researchers. Another avenue is to learn from the methods applied by the studies reviewed in this chapter. Both theoretical and methodological developments are probably also required due to the built-in pro-innovation bias in innovation diffusion theory. Several theoretical contributions are identified in table 4.3, and we have also developed and shown an application of a possible analytical framework in this study. Undesirable consequences of innovation are notoriously complex, multifaceted, and multilevel, and they transcend artificially created scientific boundaries. They require multidisciplinary research, cross-boundary dialogue, and often also a longitudinal approach. They are difficult, but not impossible to study, as attested by scholars in this book.

NOTES

1. In this chapter, 'undesirable consequences' are used as a collective term including all possible synonyms of the two words we have found during our research, such as 'negative effects' (see tables 4.1 and 4.4). In the interest of legibility and to reduce repetitiveness, we also use the synonyms in the text.
2. The results of the first article search were presented in a conference paper at the XX ISPIM conference, The Future of Innovation, Vienna, June 21–24, 2009. The authors acknowledge with gratitude assistance with data gathering by Alexander Aminoff and Andreas Eriksson.
3. The second search with expanded search terms and focused on 'innovation studies' was conducted August 2011; see later in this chapter for more information.
4. Merton's (1936) terminology is to some degree superseded, but the classification remains unchallenged.
5. The choice of search terms was initially the same as the one published in *The Oxford handbook of innovation* (2005, p. 2), which searched only articles with 'innovation' in the title, but it was expanded to catch more articles.
6. The five most influential journals are: *Research Policy*, *Industrial and Corporate Change*, *Journal of Evolutionary Economics*, *Economics of Innovation and New Technology*, and *Structural Change and Economic Dynamics*. The only journal to have published a special issue with critical perspectives on innovation is not among the five: *Technology Analysis & Strategic Management*, 2000, 12(3).

REFERENCES

*Refers to articles identified in the two article searches.

*Abrahamson, E. (1991). Managerial fads and fashions: The diffusion and refection of innovations. *Academy of Management Review*, 16, 586–612.

*Alam, I. (2003). Innovation strategy, process and performance in the commercial banking industry. *Journal of Marketing Management*, 19, 973–999.

Andersen, S. C., & Serritzlew, S. (2007). The unintended effects of private school competition. *Journal of Public Administration Research & Theory*, 17, 335–356.

Anderson, N., De Dreu, C. K W., & Nijstad, B. A. (2004). The routinization of innovation research: A constructively critical review of the state-of-the-science. *Journal of Organizational Behavior*, 25, 147–173.

*Aubert, P., Caroli, E., & Roger, M. (2006). New technologies, organization and age: Firm-level evidence. *Economic Journal*, 116, 73–93.

Baert, P. (1991). Unintended consequences: A typology and examples. *International Sociology*, 6, 201–210.

Barry, N. (1982).The tradition of spontaneous order. Retrieved February 22, 2009, from the Homepage of Library of Economics and Liberty: http://www.econlib.org/library/Essays/LtrLbrty/bryTSO.html

Binbasioglu, M., & Winston, E. (2004). Systems thinking for identifying unintended consequences of IT: Packaged software implementation in small businesses. *Journal of Computer Information Systems*, 45, 86–93.

*Calantone, R., Chan, K., & Cui, A. (2006). Decomposing product innovativeness and its effects on new product success. *Journal of Product Innovation Management*, 23, 408–421.

Calfee, J. E. (1987). Cigarette advertising regulation today: Unintended consequences and missed opportunities? *Advances in Consumer Research*,14, 264–268.

*Chigona, W., & Licker, P. (2008). Using diffusion of innovations framework to explain communal computing facilities adoption among the urban poor. *Information Technologies & International Development*, 4, 57–73.

*Clark, P., Bennett, D., Burcher, P., Newell, S., Swan, J., & Sharifi, S. (1992). The decision-episode framework and computer-aided production management. *International Studies of Management & Organization*, 22, 69–80.

*Desmarchelier, P., & Szabo, E. (2008). Innovation, food safety and regulation. *Innovation: Management, Policy & Practice*, 10, 121–131.

Dodgson, M., & Rothwell, R. (Eds.). (1996). *The handbook of industrial innovation*. Cheltenham: Edward Elgar.

Dosi, G. (1988). *Technical change and economic theory*. London: Pinter, 1988.

*Evangelista, R., & Savona, M. (2002). The Impact of Innovation on Employment in Services: evidence from Italy. *International Review of Applied Economics*, 16(3), 309–318.

*Evangelista, R., & Savona, M. (2003). Innovation, employment and skills in services: Firm and sectoral evidence. *Structural Change & Economic Dynamics*, 14, 449–474.

Fagerberg, J. (2010, May). *Exploring the knowledge base of innovation*. Presentation at Beyond the Pro-Innovation Bias workshop, Hanken School of Economics, Helsinki.

Fagerberg, J., Mowery, D. C., & Nelson, R. R. (2006). *The Oxford handbook of innovation*. Oxford: Oxford University Press.

Fagerberg, J., and Verspagen, B. (2009). Innovation studies: The emerging structure of a new scientific field. *Research Policy,* 38, 218–233.

Fairhurst, G. T., Cooren, F., & Cahill, D. J. (2002). Discursiveness, contradiction, and unintended consequences in successive downsizings. *Management Communication Quarterly,*15, 501–540.

*Farrell, K., & Runge, C. (1983). Institutional innovation and technical change in American agriculture: The role of the New Deal. *American Journal of Agricultural Economics*, 65(5), 1168–1173.

Feller, I., & Menzel, D. C. (1977). Diffusion milieus as a focus of research on innovation in the public sector. *Policy Sciences,* 8, 49–68.

*Fichman, R. G. (2004). Going beyond the dominant paradigm for information technology innovation research: Emerging concepts and methods. *Journal of the Association for Information Systems,5*, 314–355.

Freeman, C. (1974). *The economics of industrial innovation*. Harmondsworth: Penguin Books.

Freeman, R. E. (1984). *Strategic management: A stakeholder approach*. Boston: Pitman.

Fry, M., & Polonsky, M. J. (2004). Examining the unintended consequences of marketing. *Journal of Business Research,* 57, 1303–1306.

German, L., Mowo, J., & Kingamkono, M. (2006). A methodology for tracking the 'fate of technological interventions in agriculture. *Agriculture and Human Values,* 23, 353–369.

*Giuri, P., Torrisi, S., & Zinovyeva, N. (2008). ICT, skills, and organizational change: Evidence from Italian manufacturing firms. *Industrial & Corporate Change*, 17(1), 29–64.

Glinavos, I. (2008). Neoliberal law: Unintended consequences of market-friendly law reforms. *Third World Quarterly,* 29, 1087–1099.

Godin, B. (in press). *Innovation and culture (Part I): The diffusion controversy and its impact on the study of innovation*. (Working Paper No. 12, Project on the Intellectual History of Innovation).

*Goeschl, T., & Swanson, T. (2000). Genetic use restriction technologies and the diffusion of yield gains to developing countries. *Journal of International Development*, 12, 1159–1178.

Gopalakrishnan, S. (2000). Unraveling the links between dimensions of innovation and organizational performance. *Journal of High Technology Management Research,* 11, 137–153.

*Gross, M. (2005). Technology development as innovative crisis: Georg Simmel's reflections on modern science and technology. *Perspectives on Global Development & Technology*, 4, 45–61.

Harrison, M. I., Koppel, R., & Bar-Lev, S. (2007). Unintended consequences of information technologies in health care: An interactive sociotechnical analysis. *Journal of the American Medical Informatics Association,*14, 542–549.

*Herring, H., & Roy, R. (2007). Technological innovation, energy efficient design and the rebound effect. *Technovation*, 27, 194–203.

Hill, W. (2003). *Innovation will occur when parties: Share gain and pain.* New York: McGraw-Hill.

Holzer, B., & Millo, Y. (2005). From risks to second-order dangers in financial markets: Unintended consequences of risk management systems. *New Political Economy,* 10, 223–245.

*Hull, R., & Kaghan, W. (2000, September). Innovation—But for whose benefit, for what purpose? *Technology Analysis & Strategic Management*, 317–325.

Irwin, J. G., Hoffman, J. J. & Geiger, S. W. (1998). The effect of technological adoption on organizational. *International Journal of Organizational Analysis (1993–2002),* 6, 50–64.

Iyengar, R. (2008). *I'd rather be hanged for a sheep than a lamb: The unintended consequences of 'three-strikes' laws* (Working Paper No. 13784). Cambridge, MA: National Bureau of Economic Research.

Johns, G. (1993). Constraints on the adoption of psychology-based personnel practices: Lessons from organizational innovation. *Personnel Psychology,* 46, 569–592.

Kitchener, M. (2002). Mobilizing the logic of managerialism in professional fields: The case of academic health centre mergers. *Organization Studies,* 23, 391–420.

*Koeller, C. (1995). Innovation, market structure and firm size: A simultaneous equations model. *Managerial & Decision Economics*, 16, 259–269.

*Koeller, C. (2001). The influence of technological opportunity on union membership in U.S. manufacturing industries. *Journal of Labor Research*, 22, 173–183.

*Kossek, E. (1989). The acceptance of human resource innovation by multiple constituencies. *Personnel Psychology*, 42, 263–281.

Legge, Jr., J. S. (1983). The unintended consequences of policy change: The effect of a restrictive abortion policy. *Administration & Society,*15, 243–256.

Lissoni, F., & Metcalfe, J. S. (1994). Diffusion of innovation ancient and modern: A review of the main themes. In M. Dodgson & R. Rothwell (Eds.), *The handbook of industrial innovation.* Cheltenham: Edward Elgar.

Lowrey, T. M. (1991). The use of diffusion theory in marketing: A qualitative approach to innovative consumer behavior. *Advances in Consumer Research,*18, 644–650.

McAulay, L. (2007). Unintended consequences of computer-mediated communications. *Behavior & Information Technology,* 26, 385–398.

McNamara, G., Moon, H., & Bromiley, P. (2002). Banking on commitment: Intended and unintended consequences of an organization's attempt to attenuate escalation of commitment. *Academy of Management Journal,* 45, 443–452.

Merton, R. K. (1936). The unanticipated consequences of purposive social action. *American Sociological Review,* 1, 894–904.

Merton, R. K. (1968). *Social theory and social structure* (Enlarged ed.). London: Collier-Macmillan.

*Nicholas, T. (2008). Does innovation cause stock market runups? Evidence from the Great Crash. *American Economic Review,* 98, 1370–1396.

Pinto, F. N. L. (1996). *Understanding the process of innovation through the theories of learning and communication* (Publication No. AAT MQ33429). Mellon Fellowship dissertation, Lakehead University, Canada.

*Ramdani, B., & Kawalek, P. (2007). SMEs and IS innovations adoption: A review and assessment of previous research. *Academia: Revista Latinoamericana de Administración*, 39, 47–70.

Robertson, M., Swan, J., & Newell, S. (1996). The role of networks in the diffusion of technological innovation. *Journal of Management Studies*, 33, 333–359.

Rogers, E. M. (1976). New product adoption and diffusion. *Journal of Consumer Research*, 2, 290–301.

Rogers, E. M. (1983). *Diffusion of innovations* (3rd ed.). New York: Free Press.

Rose, D. R., Clear, T. R., & Ryder, J. A. (2001). Addressing the unintended consequences of incarceration through community-oriented services at the neighborhood level. *Corrections Management Quarterly*, 5, 62–70.

Saha, A. (1998). Technological innovation and Western values. *Technology in Society*, 20, 499–504.

*Schroeder, D. (1990). A dynamic perspective on the impact of process innovation upon competitive strategies. *Strategic Management Journal*, 11, 25–41.

Schumpeter, J. (1939). *Business cycles: A theoretical, historical and statistical analysis of the capitalist process* (1st ed.). New York: McGraw-Hill.

*Scranton, P. (2006). Technology, science and American innovation. *Business History*, 48, 311–331.

Shavinina, L. V. (Ed.). (2003). *International handbook on innovation.* Amsterdam: Elsevier.

*Skuras, D., Tsegenidi, K., & Tsekouras, K. (2008). Product innovation and the decision to invest in fixed capital assets: Evidence from an SME survey in six European Union member states. *Research Policy*, 37, 1778–1789.

Smith, A. (1789). *An inquiry into the nature and causes of the wealth of nations* (5th ed.). London: Strahan and Cadell.

*Smith, D., & Skalnik, J. (2003). Biotechnology in the agricultural sector: A challenge to consumer welfare. *International Journal of Consumer Studies*, 27, 277–282.

*Smoller, M. M., Osborn, R., & Price, K. (2001). Purposeful unintended consequences with innovative derivatives. *Thunderbird International Business Review*, 43, 433–449.

Stoneman, P. (1995). *Handbook of the economics of innovation and technological change*. Oxford: Blackwell.

*Swanson, E., & Ramiller, N. (2004). Innovating mindfully with information technology. *MIS Quarterly*, 28, 553–583.

*Szulanski, G., & Jensen, R. (2008). Growing through copying: The negative consequences of innovation on franchise network growth. *Research Policy*, 37, 1732–1741.

Tidd, J., Bessant, J., & Pavitt, K. (2005). *Managing innovation: Integrating technological, market, and organizational change.* New York: Wiley.

White, B. J., & Ramsey, V. J. (1978). Some unintended consequences of top down organization development. *Human resource management*, 17, 7–14.

*Winkler, R. (2005). Structural change with joint production of consumption and environmental pollution: A neo-Austrian approach. *Structural Change & Economic Dynamics*, 16(1), 111–135.

Part II

Understanding the Systemic Nature of Innovation

5 Accelerating the Innovation Race
Do We Need Reflexive Brakes?

Mervi Hasu, Karl-Heinz Leitner, Nikodemus Solitander, and Urmas Varblane

> When running among fast runners, to slow down means to be left behind; when running on thin ice, slowing down means the real threat of being drowned. Speed, therefore, climbs to the top of the list of survival values. Speed, however, is not conductive to thinking, not to thinking far ahead, to long-term thinking at any rate . . . in the absence of thought, the skating on thin ice which is the fate of fragile individuals in the porous world may well be mistaken for their destiny. (Bauman 2000, p. 210)

This chapter analyzes the factors that contribute to the ever-accelerating innovation process and provides some tools for critical reflexivity[1] on the consequences of this 'innovation race' for the economy, society, and nature. The message of the dominant innovation discourse is that everyone needs to take part in the innovation process—all the time: at work and at home, if not as a worker who is putting in working hours at home, then as a consumer who is 'co-creating' (Prahalad & Ramaswamy, 2004) and testing out new products for the firm through processes of 'social innovation' (McElroy, 2002), at ever-increasing speeds. In innovation policy and research, time serves as a basis for a sense of constant urgency; we now live in a 'hurried culture' (Bertman, 2000). Innovations are portrayed as occurring at an accelerating pace, necessitating both fast management subjects (Thrift, 2005) and fast policy responses. The metaphor of a 'race' is often utilized in innovation policy documents—the idea is that a race is underway not only between firms, but between Western regions, nations, and cities, as well as between industrially developed regions and so-called laggard or catching-up regions. Such rhetoric is used by policy makers, intermediaries, and industrialists to mobilize action and legitimatize policy actions and investments in research aimed at innovation.

The speed imperative of innovation discourse is effectively facilitated and diffused by academic research, consulting and policy-making, which jointly reinforce each other in facilitating development (Thrift, 2005; Bryson, 2000). These actors are all central to the functioning and acceleration of 'the innovation race' as they (1) produce normative innovation theories (through research), (2) diffuse and establish them as practice (through consulting), and (3) govern them (through policy making). The speed imperative also governs these actors, as they are under constant pressure to produce more theories and models at an ever-increasing pace, diffuse them faster, establish them, and then discard them for something new.

By innovation speed we refer to 'the rate at which discoveries are converted into rent-producing assets,' that is, new products and services (Markman, Gianiodisa, Phanb, & Balkin, 2005, p. 1059), rather than the diffusion of innovations among populations (Rogers, 1983). Understanding the speed imperative and the lack of reflexivity within innovation research in terms of its adverse effects requires us to root the discussion in the evolution of modernity, as it is a deeply historical process. In innovation discourse, there is little time for anything to solidify, and there is no time for any policy or theory to serve as a point of reference for sustainable development. Little, if anything, is *sustained,* as everything moves at an ever-increasing speed. As Bauman (2000) reminds us, whereas speed in all its forms has climbed to the top of the value pyramid of modern society, it is not conductive to long-term thinking; deeper learning often requires time (Argyris, 1999).

We also refer to speed and related notions (e.g., pace, acceleration, race) not only discursively but also as a more mundane practice. In innovation as practice, the speeding up of work processes as part of the continuous introduction of new organizational routines, projects, and information systems may endanger communities' ability to critically reflect on their innovative activities and ensuing consequences. Today's capitalism draws workers into a system that is too fluid and mobile for the desire to do something well for its own sake. Flexible, global organizations do not look for craftsmanship, because getting deeply involved in any problem would be dysfunctional, because projects can end as abruptly as they begin (Sennett, 2006).

In this chapter, we explore the dynamics and possible impacts of the speed imperative in the dominant innovation discourse and its practice, thus examining how and why the speed imperative is seemingly so uncritically accepted, while also asking how innovation discourse and practice could address the question of 'slowing down' innovation. We posit that there is a growing need for *reflexivity* in relation to the steady increase of the pace of innovations itself, as well as for greater production of innovation imperatives that produce images of a constant state of urgency, which in turn necessitates fast decisions based on simple, generic models, be they in the form of policies or management theories and models.

In general terms, reflection involves giving order to situations, whereas reflexivity entails unsettling conventional practice (Cunliffe & Easterby-Smith, 2004). Cunliffe (2009, p. 98) makes a distinction between: (1) self-reflexivity, that is, the recognition that we shape and are shaped by our social experience, involving a self-dialogue on our values and assumptions, questioning of core beliefs—for instance, that we as innovation researchers play our part in accelerating the innovation race; and (2) critical reflexivity, which means the examination and unsettling of our own assumptions, actions, and their impact, that is, in this case, problematizing the adverse and unintended consequences of accelerating innovation.

In our view, the speed imperative is contradictory: On the one hand, the acceleration of innovation is socially desirable; it may help developing regions to advance and become more prosperous, and address social and environmental challenges. On the other hand, the speed imperative is a mechanism of greed and unsustainability; speeding up the introduction of new product variations into markets enables global enterprises to make more profit. However, only a fraction of new technologies and product innovations facilitate the solving of social and environmental problems, or contribute to well-being and health. Some even have demonstrably negative social and environmental consequences. The pace of innovation itself may cause new types of problems: it may negatively influence mental health, increase the dependency of society on technologies or single firms, and waste natural resources. From the point of view of the global capitalists, consumers are not supposed to question the increasing speed of consumption and the launching of "new user needs" (Ritzer & Jurgenson, 2009). Global capital benefits from low reflexivity and the low criticality of innovation systems. Therefore, we are interested in the unintended and undesirable consequences of the acceleration of innovation (discourse and practice), in terms of what it adds to or takes away from a system that may seriously inhibit the critical reflexivity of the participating communities.

To examine the subject, we identify three literature streams and theoretical debates/ideas that are particularly useful in exploring the acceleration dynamics of innovation within innovation systems and processes, as well as their unintended effects. In doing so, we use three perspectives that we consider particularly relevant: (1) the concept and idea of liquid modernity (Bauman, 2000), (2) the literature on the social shaping of technology and technology assessment (Schot, 2001), and (3) the learning literature (e.g., Argyris & Schön, 1978). These three lines of research are rooted in sociology and organizational theory and deliver perspectives on why systems or organizations often fail to critically reflect on their own development. We chose these three because each of them open up different system levels of acceleration dynamics, while simultaneously complementing each other: The global cultural/economic level (macro), national/regional innovation system and the policy level (meso), and the organizational/innovation process level. We will provide the main arguments, expand on the debates, and compare findings by discussing the opportunities of developing an integrated reflexivity framework for critically assessing the systemic consequences of acceleration, the main manifestation of the speed imperative.

CRITIQUE OF LIQUID MODERNITY: THE SPEED IMPERATIVE AND THE LACK OF REFLEXIVITY IN INNOVATION STUDIES AND INNOVATION PRACTICE

Today, the dominant view of innovation is that it is something inherently and unquestionably good—not least because it is mainly understood in relation to economic growth and competitiveness, following the Schumpeterian tradition (Chapman, 2007; van Griethuysen, 2010). Innovation research has been instrumental in diffusing the truism of 'innovate or die' (Jackson, 2009). The process of 'creative destruction' is at the heart of the dominant innovation discourse, which leads to practice in which firms continually produce novelty—not only cheaper, but also newer and more exciting products to replace the latest innovation, and all at an increasingly faster pace. Yet the adverse effects of innovation speed are consciously underrepresented in innovation literature (Kessler & Chakrabarti, 1996; Markman et al., 2005; Schmidt, 2008). What presently exists is an unrelented, technoscientific belief that new innovations will always solve the problems caused by other innovations or the innovation process (Schmidt, 2008, p. 2). If not skeptical about the goodness of innovations, we are at least worried about the lack of reflexivity within innovation studies on the speed imperative.

Bauman (2000, p. 14) compares the earlier stage of modernity to a new emerging form of 'liquid modernity.' Whereas in modernization's earlier form, the captains of industry would produce massive, bulky factories, railroads, and oilrigs to be owned and marveled at for an 'eternity,' in contrast, in liquid modernity, there is no regret on the part of the captain of industry when parting with possessions he took pride in yesterday. Instead 'it is the mind-boggling speed of circulation, of recycling, ageing, dumping and replacement' (p. 14) that produces profit, rather than the durability or reliability of the product. Herein, the role of industrialists such as the long-term president and chairman of General Motors, Alfred Sloan, and his innovations within marketing should be emphasized (Lash, 1985; Bauman, 2000), alongside Schumpeter's gales of creative destruction. Sloan was instrumental in legitimizing and diffusing the managerial innovation of planned obsolescence (Dauvergne, 2008; Usui, 2008), wherein new product designs make old ones obsolete. Whereas today normalized and naturalized as a practice, when the idea was initially developed around 1920, it was heavily criticized and rejected on moral grounds (Usui, 2008, p. 98). However, the kind of reflexivity and long-term thinking needed to address moral and ethical issues (beyond pasting it on as an argument for selling more) does not naturally fit with the discursive confines of the speed imperative. As long as people feel that 'skating on thin ice' (Bauman, 2000, p. 209) is their destiny, they will maintain a high speed in whatever they do. Deep learning, however, calls for taking one's time to recapitulate the steps taken and to understand the long-term effects of innovations; effects that span not only economic value but social and environmental effects as well. Yet we, as workers and consumers, are faced with the fact that deeper thinking about the (unintended) consequences of innovation takes 'one's mind away from the task at hand, which is always running and keeping speed whatever else it may be' (Bauman, 2000, p. 210).

HOW THE SPEED IMPERATIVE OF INNOVATION MANIFESTS ITSELF IN PRACTICE

Global business goals affect organizational temporality, speeding up product cycles, and release and response times to various design tasks. Today, the goal of making short-term profits is a strong reason for organizing and managing research and design (R&D) work effectively to meet business goals, while simultaneously maintaining the speed imperative for innovation. However, short-term business goals and long-term sustainable innovation and creativity are difficult to combine in practice. The constant decrease of release times of new products onto the market suggests that the organizing of R&D practices and schedules is increasingly intensified in private companies. On a macrolevel, R&D management seems to have succeeded in combining the effectiveness and creativity of R&D work. At the same time, individual expert work in general has been intensified (e.g., Perlow, 1999; Brown & Eisenhardt, 1997). Many branches of expert work, in particular software design, have turned from high-rank creative work into insecure, time-pressured subcontractor work.

In the following two subsections we will illustrate the previously mentioned development using two examples.

Speed Imperative, Labor, and Project Work

The most common form of organizing applied research work and innovation endeavors is commonly known as a project. Originally, the project as a unit of managing work was mainly used in the construction and manufacturing context (Cicmil, Hodgson, Lindgren, & Packendorff, 2009). By the 1990s, however, project-based organizing had been adopted as a widespread solution for contemporary organizations to develop and integrate knowledge across diverse functions for the benefit of customers (Cicmil & Hodgson, 2006). Turner and colleagues (2008) researched how project work influenced worker well-being in terms of their work/life balance and feelings of uncertainty in the temporary-based work environment, as well as how project assignments were linked to career development. Their results indicate that the most significant problems affecting employee well-being and avoiding excessive levels of work occur in small- to medium-sized projects (typically of three to nine months' duration). Larger projects guarantee easier opportunities to plan the employee's workload. A Swedish research team (Zika-Viktorsson, Sundström, & Engwall, 2006) examined the psychosocial aspects of work in multiproject settings. Their results supported earlier studies by showing that participation in multiple projects at the same time increased the risk of excessive workload, because of the lack of opportunities for recuperation between intense work periods and employees' division into multiple project roles. Project overload was also associated with impaired performance, high levels of stress reactions, and decreased competence development.

The innovation speed imperative, in terms of the ultimate respect of customers, creates unrealistically short time estimates for projects. Lindgren and Packendorff (2006) argue that current project work practices imply reproduction of masculinities such as rationality, efficiency, control, devotion of work, and strict deadlines, whereas feminization, referring to such values as teamwork, caring, empathy, interdependence, is found in the rhetoric of the overall organizational context. This discrepancy is hardly noticed in organizations, because what matters in the workplace is mainly successful project delivery, regardless of what it takes. It may be expected that such things as deadline stress at the end of a project become institutionalized. The conclusion is that as long as people accept demands on cost and time efficiency in project work, the current masculinization of project work practices will persist. What the researchers suggest is that future research should examine both what projects do to people and what people do to projects (Lindgren & Packendorff, 2006.)

The Speed Imperative and the Blurring of Boundaries Between Work and Leisure

Today, long working hours are typical worldwide (e.g., Vaguer & Van Bastelaer, 2004). However, constant disturbance-load and disruptions in office work are a daily experience in expert work. It is common that the most demanding and creative work tasks are increasingly accomplished outside regular working hours (Florida, 2002; Perlow, 1999). The blurring of boundaries between work and leisure is widespread among highly educated experts, workers have to be 'nurtured' also during (unpaid) leisure time, so that the firm can at all times and when needed gain access to the individual's epistemological resources of knowledge and creativity—all with little critical reflection (Fougère & Solitander, 2010; Thrift, 2005). However, if burnout occurs, energy turns into exhaustion, involvement into cynicism, and efficacy into ineffectiveness (Schaufeli & Bakker, 2004).

Excessively long working hours is one of the effects of the speed imperative of the innovation era. There is epidemiological evidence of ill-health due to working long hours in the form of sleeping problems, and problems in cognitive functioning and coronary health. It appears to be a risk factor for the development of shortened sleeping hours and difficulty falling asleep (Virtanen, Ferrie, et al., 2009). Long working hours may also have a negative effect on cognitive performance in middle age (Virtanen, Singh-Manoux, et al., 2009). Overtime work is related to an increased risk of incident coronary heart disease (CHD), independent of conventional risk factors: overtime work adversely affects coronary health (Virtanen et al., 2010). This suggests that problems exist on both micro- and macrolevels. Interestingly enough, employee health and work-related well-being, motivation, and engagement have become a major concern for the human resources function of the most innovative companies worldwide. However, innovation research has not shown much interest in the potential negative impacts of the temporal pacing of innovation-related work, such as project overload and continuous long working hours.

As previously discussed, the speed imperative of the innovation era has multiple unintended and undesirable consequences for individuals, work communities, work processes, and also—potentially—for the products and services that they design, produce, and implement. The previously cited observations mean that putting workers under excessive time pressure and work overload not only jeopardizes their well-being, but places the quality of work and long-term development under risk (e.g., Zika-Viktorsson et al., 2006). In addition, we posit that the acceleration of innovation, in terms of increasing disruptions at work and the growing speed of work processes, at the same time leave no time or opportunity to critically reflect on what is happening now and what will follow.

In our view, low reflexivity in innovation practices seems to be compatible with low reflexivity in the dominant paradigm of innovation studies. Innovation practitioners and entrepreneurs in the field may be too busy or exhausted to genuinely interact with policy makers and researchers and provide unpolished accounts of innovation practices. Mainstream innovation scholars themselves seldom, if ever, engage in intensive fieldwork to explore innovation 'in the flesh.' The speed imperative as a simple rhetoric may be beneficial for many political and official institutions, even for the practitioners themselves. Consequently, the speed imperative and its lack of reflexivity travels: It is well-received rhetoric in new contexts.

We will next discuss the seemingly easy transferability of the innovation speed imperative from one context to another, by providing two examples.

Diffusion of the Innovation Speed Imperative from R&D-Intensive Industries to Low-Technology Industries or Transitional Economies

Oversimplification of the importance of the development of science-intensive industries and thus the role of expenditures in R&D prevails in the E.U., as in other developed economies. In the competition between countries concerning who is able to increase R&D expenditures to the highest level from their GDP, countries have forgotten that this is an input target only (see Archibugi & Coco, 2005). Research and development expenditures say very little about how efficiently these expenditures will be made.

If not accompanied by sensible policy measures, there is the risk of waste—spending money unwisely. Particularly in the new E.U. member states, the proposed rapid increase of R&D expenditures up to the unrealistically high level of 3% already in 2013, without significant reforms in the structure of R&D spending, would be a serious waste of limited resources (Varblane, Dyker, Tamm, & von Tunzelmann, 2007).

Diffusion of the Innovation Speed Imperative from the Private to the Public Sector

The public sector has now also entered the innovation imperative era. It faces the simultaneous challenge of cost-efficient service production while assuring the quality of new services at the microlevel. The introduction of new computer-based services and management tools is now becoming everyday life in public sector organizations. Work in the public sector is increasingly influenced by the ideas of the new public management and the ethos of the "New Economy" (Thrift, 2005). Business consultants, performance measurement, and enterprise resources planning (ERP) systems have also entered public sector organizations. For instance in Finland, major public funding organization TEKES, and also SITRA, one of the key players in innovation policy, are both increasingly targeting public sector innovation for research funding and supportive measures. In Europe, the innovation imperative and the sense of urgency in innovation is likely to spread rapidly from the business innovation context into the public sector and service contexts such as social and welfare services, health care, and elderly care. Empirical evidence concerning the effects of this development on service quality is still scarce.

The previously mentioned examples also underline another reflexivity gap within innovation studies; that of a lack of a field of innovation ethics (Schmidt, 2008). Although we have a field of engineering ethics, the link to innovation studies is implicit at best (ibid.). Establishing a field of innovation ethics would force such reflexivity from the periphery toward the center.

We will next turn to the literature streams, which offer approaches to reduce the negative aspects of technological development and innovation.

HOW TO CREATE REFLEXIVITY CAPABILITIES IN AN INNOVATION SYSTEM

The traditional innovation literature, in particular the innovation systems literature, offers less debate on the negative and unintended effects of innovation. The exception is the literature stream dealing with technology assessment (TA) and the social shaping of technology (SST). Here it is assumed that participation by many innovation system actors along the entire innovation process balances the different economic, social, and environmental impacts of innovation and minimizes any negative impacts. The quasi-evolutionary model developed by Schot (1992), for instance, was one of the most prominent attempts in this respect and laid the foundation for further efforts to bring these insights to bear for the constructive purposes of managing sociotechnical change. The aim of his approach is to shape and influence the processes of technological change with as few negative social and environmental impacts as possible.

Strategic Niche Management Builds Abilities to Consider Alternative Pathways and Visions among Network Actors

In a similar vein, the strategic niche management (SNM) approach was developed to understand and subsequently guide efforts to establish new emerging technologies that have the potential of contributing to more sustainable development (Hoogma, Kemp, Schot, & Truffer, 2002). SNM stresses the need to nurture new emerging technologies that are not compatible with the prevailing technological regime in protected niches, to enable first- and second-order learning processes with the technology in its social context, until it is sufficiently stable and robust to compete with dominant technologies under competitive market conditions. This development is associated with the hope that the new technology represents a more sustainable technological regime, which considers negative impacts on social life and the environment and can subsequently contribute to either replacing or transforming the dominant paradigm. However, in this sense, such an approach still considers that specific types of innovations are needed to transform systems and create markets with less negative impacts on society or the environment, and hence innovation ultimately has a positive connotation in this context.

The previously mentioned approaches often call for a high level of participation for the co-development and shaping of technological development, also triggered by constructing common visions for future development. SST aims at developing images of and pathways toward the future, as well as at specifying the elements and working configurations that are relevant at the different levels of a transition process toward a newly emerging, more sustainable paradigm. Being set up as participatory processes involving different actors and stakeholders, these processes contribute to mobilizing actors at the different levels of action on the individual, micro-, meso-, and macrolevel, as well as to enhancing connectivity and capability for interactive learning and to informing the strategies and action plans of individual actors. Voß and colleagues (2006) more recently called for 'reflexive governance' with respect to mastering the transformation of a system toward more sustainability, and Eriksson and Weber (2007) have argued for the necessity of 'adaptive foresight' when critically constructing, debating, and shaping future technological development.

Kuhlmann and Smits (2004) have argued that such approaches require the specific 'strategic intelligence' of a system that may allow reflection on the negative effects of uncontrolled technological developments (Barre, 2001). TA and SST approaches have, for instance, been used in technology and in foresight studies for defining R&D priorities or impact assessment, for example, for nanotechnology, but have so far rarely been adopted to more generically address the negative effects of the increasing pace of innovations.

Need for Critical Reflexivity among Intermediary Institutions and Academics

As regards the growth of the 'innovate or die' and speed of innovation paradigms, since the 1970s, there has also been a rapid increase in governmental regulations targeted at the environmental and health risks connected to innovations. New technologies in particular have been increasingly scrutinized, monitored, and regulated (Schot, 2001). Beck has argued in this context that we have entered a reflexive modernization in which society increasingly confronts its own problems (Beck, 1992). Reflexivity here refers to the 'ability to encounter the familiar as new (unfamiliar) [which helps] to reflect on one's reflections' (Antonacopoulou, 2010, p. 7). However, what is largely missing from innovation research today is critical reflexivity, not least regarding the overt positive connotation of innovation, a problematization that is the theme of this book. As previously discussed, one reason for the lack of reflexivity within innovation studies is the prominence of technocentric thinking, in which there is a strong belief that problematizations are of less importance, because (1) innovations are benevolently good, and (2) technology will eventually solve all adverse effects created by mankind (Feenberg, 1999; Schmidt, 2008). Another reason for the difficulty in being reflexive is that innovation research is part of the reality that it depicts.

Academic disciplines, such as innovation studies themselves, are important in the legitimization of certain understandings as facts, making them uncontestable and unquestionable. Academia is a central part of what Thrift (2005) calls the cultural circuit of capitalism. In modern forms of capitalism, the production and exchange of knowledge is increasingly managed 'through the production of theories, texts, and practical protocols such as seminars that stabilize them, embody them in subjects' (p. 46). Thrift (pp. 45–46) identifies the forms of soft capitalism by which he means, for instance, explicit social engineering for maximizing interaction and thereby maximizing learning and innovation. Examples of this discursive apparatus are intermediary institutions, such as consultants, management gurus, business schools, and agencies for innovation systems (such as VINNOVA in Sweden, SITRA in Finland, ESTAG in Estonia, and FFG in Austria). As academia is inherently part of the reality it depicts, there cannot be a one-way correspondence between theoretical category and innovation reality, because the innovation process partly emerges from the act of theorizing about it (Barnes, 1992). Another contributing and linked fact to the lack of reflexivity is that academia itself is part of the same process of liquid modernity as described in the previous section. The 'innovate or die' mentality is mirrored in the idea of 'publish or perish' (Meriläinen, Tienari, Thomas, & Davies 2008; De Rond & Miller, 2005), wherein academic 'innovations' are produced and reproduced before they have an opportunity to solidify. This all adds up to a milieu that is not susceptible to an atmosphere of critical reflexivity.

Need for Critical Reflexivity among Policy Makers

To slow down the innovation race, there must be an increase in reflexivity within innovation studies and innovation policy in particular (Schmidt, 2008). Thereby a lack of reflexivity may be called "reflexivity failure," referring to the notion of systems failures as often used to legitimize innovation policy interventions (Woolthuis, Lankhuizen, & Gilsing, 2005; Weber, 2006). Innovation systems that have a lack of such critical capacities hence reveal a reflexivity failure.

The European Commission, for instance, has started in recent years to reflect more critically on the impacts of science, technology, and innovation, and to call for more responsible research and innovation (Felt et al., 2007). The Commission's expert group consequently argues for the introduction of structured ways of appraising the projected benefits of innovation by involving citizens more actively in science and innovation processes. The expert group argues: *"An important change in the governance of innovation would be strategic development or improved European institutional capacity to deliberate and resolve normative questions concerning the prior shaping of science and innovation over their directions as well as their scale and speed"* (Felt et al., 2007, p. 11). It is interesting that in this report some experts, for the first time, expressed their criticism of the Lisbon Agenda and the aim to achieve as much innovation as possible, as fast as possible. We should not " . . . *just be about the more and the faster the better*" (p. 10).

The expert group organized by the commission is also more critical toward the "speed imperative" and reveals that a specific rhetoric is used to mobilize all levels of society to increase innovation, as Europe will otherwise "fall behind." The expert group further critically reflects on this development and the current policy debates, and argues: *'The speed imperative, to act immediately, "before it is too late," is tightly linked to European S&T policy. A January 2003 EC press release warns: "Today's fundamental research will turn into tomorrow's growth, competitiveness and better quality of life. The US has understood this. The EU is still lagging behind. Ours is a wake-up call: we need to act now to reverse this situation and fill the gap." Within this temporal narrative, reinforcing the singular, narrow, 'no-alternative' vision of scientific progress, social mechanisms such as law and ethics, or public engagement of a serious kind, are always defined as lagging or braking progress'* (Felt et al., 2007, p. 77).

The expert group is aware that such a change of the dominant logic is difficult to achieve and would require an institutional change. A more critical public reflection and deliberation is needed to change conventional understandings and the development of more diversified images about future development. New ways are required to promote 'reflexive thinking,' and society and innovation systems require new institutions and procedures to achieve more inclusive and pluralistic discussion and learning, involving all actors of the innovation system. By establishing new and deliberative forms of collective experimentation (e.g., Stirling, 2008), we can strengthen these "reflexive capacities" within society (Felt et al., 2007). New forms of innovation, such as open innovation, user-led innovation, open source software development, and community innovation, are considered promising new innovation models that reveal the emergence of a new regime of collective experimentation. However, more interestingly, the authors of the commission report assume that these new forms of distributed innovation will even lead to a slowing down of the innovation process: As the processes are typically loosely coordinated, and as the users and other stakeholders have their own contexts and logics to consider, the process becomes slow. This leads, inspired by the 'slow food' movement, to the proclamation of a 'slow innovation' program (Felt et al., 2007, p. 27).

The current innovation policy of the European Commission to address Grand Challenges can already be interpreted as a sign for a shift within the policy domain to steer scientific and technological development more explicitly toward the real needs of society. Moreover, within the impact assessment exercise, the Commission has started in recent years not only to assess and discuss the positive impacts of publicly funded research and innovation but also the negative and unintended consequences. Schomberg (2011) calls in this context for a quest for the "right" impacts of science and technology and presents a framework for responsible research and innovation. Like other authors already mentioned, Schomberg (2011) promotes the use of participatory foresight exercises that keep alternative developments in sight for possible public policy responses and for enabling democratic choices at early stages of technological development. Participatory exercises and initiatives in the area of nanotechnology or genetically modified organisms organized in the last years are examples of a deliberative approach of reflexive and critical technological development, considering issues of risk, ethics, and sustainability more comprehensively and hence enabling learning on the systems level.

Moreover, the recent shaking up of society in the course of the Fukushima nuclear power plant accident in Japan due to a serious earthquake is a clear sign that such development has already become more dynamic, and that society expresses its voice in the direction of future technological development.

We will next address the question of how more critical learning about the direction and speed of the innovation process could take place on the organizational level. We will discuss the organizational learning perspective's opportunity to slow down innovation processes.

HOW TO CREATE REFLEXIVITY CAPABILITIES IN ORGANIZATIONAL INNOVATION PROCESSES: FROM ACCELERATING LEARNING TO REFLEXIVE LEARNING

Learning is positively associated with the acceleration of innovation (Eisenhardt & Tabrizi, 1995). Learning, collaborative learning in particular, in terms of the individuals' desire to reach beyond their known capabilities together to create (John-Steiner, 2000), is a powerful asset and driver for innovation. Learning in itself is supposed to be reflexive (Schön, 1987). It is difficult to imagine that the individual desire and organizational patterns for accelerating learning in relation to innovation might have undesirable and unintended consequences. However, as discussed previously in this chapter, we are witnessing the constant need for speed not only in innovation but also in learning in contemporary organizations. This development may not be entirely positive.

How, in the first place, has learning in innovation been conceptualized and utilized in innovation studies? When and how did the first signs of reflexive learning in innovation, that is, learning that is not totally 'black box' or 'automatic,' 'uncritical,' or 'uncontextual' emerge in innovation studies? Is the learning perspective advanced within the dominant innovation discourse, or is the perspective being developed outside the dominant view?

Lack of Reflexivity Within Conventional Learning Approaches in Innovation Studies

Widely known innovation researchers Mark Dodgson and Eric von Hippel paved the way for conceptualizing learning in connection to actual organizational environments and empirical findings.[2] Dodgson (1991; 1993; 1996) studied technological collaboration, especially learning, trust, and interfirm technological linkages, and extended the technological perspective toward an organizational and managerial view. Dodgson (1991) argued that learning occurs through purposive activity, and not automatically. According to Dodgson, learning can be seen to have occurred when organizations perform in changed and improved ways, and when competences are better defined, more appropriate, and effectively implemented. Dodgson (1996) later extended the notion of learning to include broader aspects of the firm's competitiveness, survival, and (strategic) development in a changing environment. Although he did not discuss this developmental aspect explicitly, his view also requires collective agency in learning. He did not, however, further elaborate on its developmental or reflexive characteristics.

Von Hippel and his colleagues first identified failures in innovation sites as a resource for learning. Von Hippel used the notion of 'sticky information' to refer to R&D practitioners' difficulties in technical problem-solving (von Hippel, 1994; 1998). Von Hippel and Tyre's (1995; 1996) work somewhat developed and opened the "black box" of learning. It seems, however, that the analysis of problem discovery in machine use was restricted to technical problems. 'Sticky' information was associated with technical complexity and the transfer of technical information. Hence, the authors did not discuss the significance of work organization, power relations, or the effects of implementation. Learning was neither an automatic by-product of "doing," nor was it explicitly addressed as proactive or reflexive. The nature of learning remained tacit.

The previously cited scholars approached learning empirically, but did not discuss learning for critically debating the effects of innovation. Innovation-related learning is seen as positive and desirable, regardless of the consequences. Learning is good because it may accelerate the adoption of innovation into the market and the realization of new user needs. In our opinion, this view has not advanced among the more recent innovation scholars of the dominant innovation research.

How, then, can organizations better achieve a sense of recognition for the various stages, challenges, participants' points of view, and effects of innovation? How is the need for reflexive learning during a transitional stage of innovation recognized and mutually understood?

Emerging Reflexive Approaches to Learning in Innovation

More reflexive learning approaches related to innovation have been discussed outside the dominant innovation paradigm. An alternative view on learning in innovation (e.g., organizational learning, Argyris & Schön, 1978; expansive learning, Engeström, 1987; innovation as strategic reflexivity, Sundbo & Fuglsang, 2002; social learning, Williams, Slack, & Stewart, 2005) would suggest that learning is seen as a purposive activity carried out by the innovation community or network to recognize and proactively evaluate, direct, and develop the communities' capabilities of mastering both the potentially positive and negative consequences of innovative activity (Hasu, 2001, p. 38). This is a challenging task to accomplish in practice. It is unlikely that practitioners alone would devote time to this kind of purposive learning activity without a clear incentive.

Private innovation consultancy firms may not wish to accomplish this interactive and practical task either, due to the fact that the increased critical reflexivity of their customers would diminish customers' dependency on the consultant firm and thereby weaken economic prospects. Researchers of social studies of science, technology, and innovation may play a more active participant role here, presenting empirical evidence on the potential negative effects of speeding up innovation, and providing social, environmental, and ethical arguments for the benefit of critical reflexive learning. Academic practices in general have not been proactive in taking action in this matter in practical work contexts. Innovation researchers have not, with a few exceptions (e.g., Hoogma et al., 2002; Miettinen & Hasu, 2002; Saari & Kallio, 2010), decisively facilitated critical and reflexive learning among the innovation network participants on the organizational level.

Slowing down too fast, or uncritical learning, may mean, for instance, that the organization- and network-practitioners discover the historical trajectory of the innovation from the point of view of different parties and participant groups. Organizations and institutions go through historical changes largely independent of the specific innovation they produce or implement. These kinds of historical changes may include broad transitions from craftwork and one-off production to automated work and mass production. Organizations go through historical changes in ways of doing things: they shift from one pattern of work or performance to another. Within the innovation process, the development and implementation of the new product is often interrelated with the organizational transitions that take place in the key organizations involved. For instance, the transformation of a specific technology into a clinical instrument may be altered by simultaneous organizational transitions in the adopter institutions (Hasu, 2001, pp. 37–41).

Organizations participating in the innovation network may benefit from a simulation experiment: Purposeful sense-making of the emerging innovation trajectory and its anticipated outcomes from different viewpoints. An innovation researcher may facilitate the organizational learning process of the practitioners by providing them with his or her data and analysis, and further discussing the potential interpretations and concerns of the effects of innovation with them (e.g., Miettinen & Hasu, 2002). Researchers may also design a particular reflexivity process together with the key innovation network participants and teach the process to the practitioners for them to independently utilize it. A case in point here is, for instance, the recently introduced developmental impact evaluation method that facilitates learning in innovation networks on the organizational level (Saari & Kallio, 2010). Developed and applied in the context of research work, this method opens up the concept of 'research impact' as a qualitative learning challenge, rather than an accountable target to be judged. It may be adopted as a continuous way of managing and renewing cross-disciplinary research programs, research teams, or networks (p. 227).

Voicing the Reflexive Brakes in Practice

If we are in the constant process of speeding up the innovation process, how can we be more reflexive about such a development and assess the proper speed and right impact of research and innovation on the macrolevel? Can we build up capacities and strategies to overcome such a development? What are barriers and challenges? How should we take action?

This call for reflexive practices concerns not only policy makers and practitioners, but also individual innovation researchers. They should make the methodologies that they use *louder*: What is seen and not seen, who is included and excluded, what activity is categorized and what is left uncategorized in various innovation research contexts? Qualitative case study evidence shows that behind seemingly successful innovation endeavors, a great deal of practical shop floor work of early adopters with messy problems, setbacks, and miscommunications remains largely invisible to researchers and policy makers (Hasu, 2000a; 2000b; 2001; 2005; Hasu & Engeström, 2000; Hyysalo, 2004; Miettinen & Hasu, 2002). Work that goes unnoticed is not represented in official innovation accounts, including informal ways of knowing (Star, 1991a; 1991b; Star & Strauss, 1999; see also Forsythe, 1999). Hence, not only practical innovation-related work but also people's individual prerequisites and incentives to participate in innovation endeavors are typically missing in conventional innovation research.

CONCLUSIONS AND DISCUSSION

We have discussed in this chapter barriers and challenges facing organizations and societies when engaging in critical reflexivity about the development and use of new technologies and innovations on the micro- and macrolevel, and have referred to three different literature streams. Our analysis has shown the challenges to and the *emerging possibilities* for building up reflexivity and "braking capabilities" within a system that is constantly speeding up its pace.

In our view, society—the innovation race—needs reflexive brakes. The literature we have synthesized suggests that broader levels of participation, new codes of conduct for research and innovation, and an institutional change are needed on the cultural, sociotechnical, and organizational level to build up reflexivity capacities.

Table 5.1 summarizes our main findings concerning the three perspectives: liquid modernity, the social shaping of technology and technology assessment, and organizational/social learning. We compare the perspectives in three evaluative respects: (1) the objectives of acceleration, (2) the context of acceleration, and (3) emerging reflexive practices.

We suggest that any one of the perspectives alone is not enough to provide us with the anticipated reflexive brakes. The complexity of the current innovation era demands multivoiced and multilevel reflexivity capabilities: The combination of the three perspectives summarized in table 5.1. Our exploratory study has shown a substantial need for more research of the dynamics of such processes as concrete practices among practitioners, policy makers, and intermediary institutions, innovation researchers, and consumers.

Table 5.1 Three System Levels for Understanding the Dynamics of Acceleration in Innovation

Levels of understanding the innovation acceleration			
	Liquid modernity	Social shaping of technology and technology assessment	Organizational/social learning perspective
1. Objectives of acceleration	Acceleration of profit-production by planned obsolescence of new product and service designs	Acceleration of co-development and shaping of technology for technological and societal problem-solving	Acceleration of development, co-creation, implementation and diffusion of new products and services into market
2. Context of acceleration	Paradigms of production and consumption in the global economy and capitalist culture	Socio-technical systems, institutions and networks related to technological paradigms (regimes)	Organizations, networks, and markets
3. Emerging reflexive practices	Critique of the consequences of liquid modernity No practical applications	Critical sustainability approaches and shaping/nurturing of technology such as strategic niche management (SNM)	Participative, practice-based learning and evaluation approaches,strategic reflexivity approach

It is important to stress that reflexivity is not meant to be comfortable for us as innovation researchers, policy makers, or innovators, as it is explicitly about the unsettling of conventional practice (Cunliffe & Easterby-Smith, 2004). Following Cunliffe (2009, p. 98), we make a distinction between: (1) self-reflexivity, that is, the recognition that we shape and are shaped by our social experience, involving a self-dialogue on our values and assumptions and questioning of our core beliefs—this is largely what we have attempted to do in the previous sections; and (2) critical reflexivity, which means the examination and unsettling of our own assumptions, actions, and their impact, that is, in this case problematizing the adverse and unintended consequences of the speed imperative. This is much harder to do in practice, and there is no silver-bullet solution in terms of how to best go about it within innovation discourse and practice.

In the light of these arguments, what are the conclusions for policy makers? As shown, there are already actors, initiatives, and even official publications (e.g., by Felt et al., 2007) that stress the point as being more critical about the speed and direction of innovation. The point is most probably not to be even more critical (as this might bear the risk of being seen as too extreme or unrealistic), but to show, based on empirical evidence and cases, the very different implications of an acceleration of innovation. During the process of innovation policy design and formulation, the long term impacts of an acceleration of innovation should be considered, for example, by launching foresight projects involving many different stakeholders to debate and shape the future.

What are the implications for business managers? Companies should take a long-term perspective of their strategies and critically ask themselves how they can build a sustaining innovation culture and capacity. Managers should ask: Does the speeding of project cycles paralyze all the primary requisites of creative work and the feeling of enjoyment from work? It takes time to build and sustain complementary, creative human relationships at work, which also protect the mental health of employees. However, companies cannot guarantee the evolution of long-term trustful relationships between people in short-term, competitive, and hectic project environments. As discussed in this chapter, the mental and cognitive ill-health of employees is becoming a major new risk to economic success.

What are the conclusions for customers and citizens? Innovations do not occur outside the realm of the consumer society. The continual production of novelty would be of little value if there was not a seemingly insatiable demand for novelty among consumers: 'The restless desire of the consumer is the perfect complement for the restless innovation of the entrepreneur. Taken together, these two self-reinforcing processes are exactly what are needed to drive growth forwards' (Jackson, 2009, p. 9). Even though it makes perfect sense for ordinary consumers, customers, and citizens to partake in innovation, naive or covertly controlled participation may also be abusive. Secretly 'forced' participation in the innovation speed imperative may be attractively built through social media (games, competitions, and surveys) by global enterprises. Customers and citizens should ask: Whose voices are heard and unheard in innovation and why? They may, however, also use social media purposefully to reflect on innovation and new products and thereby influence its direction.

What are the conclusions for researchers? We offered some arguments and perspectives for more critically investigating the innovation process and its different positive and negative outcomes. We call to elaborate and extend the research agenda of traditional innovation research and thus also to leave behind the current innovation paradigm. As already stressed by Kuhn (1996), the limits to addressing new research questions and adopting new approaches due to the fact that the trajectories and paths are too strong is an inherent feature of the science system and thus a permanent challenge for any researcher.

As an attempt to create a debate, we conclude by presenting a simple 'template' for engaging in deeper reflexivity within innovation studies and practice in the form of '76 Reasonable Questions to Ask about any Innovation' (see appendix 5.1) as formulated by Jacques Ellul, a well-known French philosopher, sociologist, and critic of today's modern technological society. It is only fitting that if we are to unsettle our own assumptions and see the all-too-familiar innovation discourse as new, we must challenge ourselves by answering the questions of one of its most ardent critics.

What all of these questions have in common is that they address the areas of life that innovations affect, sometimes adversely (yet unintentionally) so, but more important, they are questions that innovation scholars often marginalize or fail to address as they do not fall within the economic domain (but instead emphasize the ecological, social, practical, moral, ethical, vocational, metaphysical, political, and aesthetic effects). Another thing that these questions have in common is that they force us to pause and to think, in the words of Bauman (2000, p. 210), to challenge the cemented belief that skating on thin ice, always running, and uncritically accepting the speed imperative is not only our fate but our destiny. These questions force us to critically reflect on the central value of liquid modernity, that '[s]olely the desiring is desirable—hardly ever its satisfaction.' (ibid, p.88)

APPENDIX 5.1

76 Reasonable Questions to ask about any innovation as formulated by Jacques Ellul[3]

Ecological

What are its effects on the health of the planet and of the person?
Does it preserve or destroy biodiversity?
Does it preserve or reduce ecosystem integrity?
What are its effects on the land?
What are its effects on wildlife?
How much and what kind of waste does it generate?
Does it incorporate the principles of ecological design?
Does it break the bond of renewal between humans and nature?
Does it preserve or reduce cultural diversity?
What is the totality of its effects, its "ecology"?

SOCIAL

Does it serve the community?
Does it empower community members?
How does it affect our perception of our needs?
Is it consistent with the creation of a communal, human economy?
What are its effects on relationships?
Does it undermine conviviality?
Does it undermine traditional forms of community?
How does it affect our way of seeing and experiencing the world?

Does it foster a diversity of forms of knowledge?
Does it build on, or contribute to, the renewal of traditional forms of knowledge?
Does it serve to commodify knowledge or relationships?
To what extent does it redefine reality?
Does it erase a sense of time and history?
What is its potential to become addictive?

Practical

What does it make?
Who does it benefit?
What is its purpose?
Where was it produced?
Where is it used?
Where must it go when it is broken or obsolete?
How expensive is it?
Can it be repaired?
By an ordinary person?

Moral

What values does its use foster?
What is gained by its use?
What are its effects beyond its utility to the individual?
What is lost in using it?
What are its effects on the least advantaged in society?

Ethical

How complicated is it?
What does it allow us to ignore?
To what extent does it distance agent from effect?
Can we assume personal or communal responsibility for its effects?
Can its effects be directly apprehended?
What ancillary technologies does it require?
What behavior might it make possible in the future?
What other technologies might it make possible?
Does it alter our sense of time and relationships in ways conducive to nihilism?

Vocational

What is its impact on craft?
Does it reduce, deaden, or enhance human creativity?

Is it the least imposing technology available for the task?
Does it replace, or does it aid, human hands and human beings?
Can it be responsive to organic circumstance?
Does it depress or enhance the quality of goods?
Does it depress or enhance the meaning of work?

Metaphysical

What aspect of the inner self does it reflect?
Does it express love?
Does it express rage?
What aspect of our past does it reflect?
Does it reflect cyclical or linear thinking?

Political

Does it concentrate or equalize power?
Does it require, or institute, a knowledge elite?
It is totalitarian?
Does it require a bureaucracy for its perpetuation?
What legal empowerments does it require?
Does it undermine traditional moral authority?
Does it require military defense?
Does it enhance, or serve, military purposes?
How does it affect warfare?
Is it massifying?
Is it consistent with the creation of a global economy?
Does it empower transnational corporations?
What kind of capital does it require?

Aesthetic

Is it ugly?
Does it cause ugliness?
What noise does it make?
What pace does it set?
How does it affect the quality of life (as distinct from the standard of living)?

NOTES

1. Reflexivity refers to the 'ability to encounter the familiar as new (unfamiliar) [which helps] to reflect on one's reflections' (Antonacopoulou, 2010, p. 7).Critical reflexivity here means the examination and unsettling of our own assumptions, actions, and their impact, that is, in this case, problematizing the adverse and unintended consequences of accelerating innovation (Cunliffe & Easterby-Smith, 2004).

2. Extended review on learning concepts in innovation studies is presented in Hasu, 2001.
3. Retrieved from http://www.thewords.com/articles/ellul76quest.htm on November 10, 2010.

REFERENCES

Antonacopoulou, E. P. (2010). Making the business school more critical: Reflexive critique based on phronsis as a foundation for impact. *British Journal of Management*, 21(1), 6–25.

Archibugi, D., & Coco, A. (2005). Is Europe becoming the most dynamic knowledge economy in the world? *Journal of Common Market Studies*, 43(3), 433–459.

Argyris, C., & Schön, D. (1978). *Organizational learning.* London: Addison-Wesley.

Argyris C. (1999) *On organizational learning.* Oxford: Blackwell

Barben, D. (2010). Analyzing acceptance politics: Towards an epistemological shift in the public understanding of science and technology. *Public Understanding of Science,* 19(3), 274–292.

Barnes, T.(1992). Redading texts of theoretical economic geography. In T. Barnes & J. Duncan (Eds.), *Writing Worlds: Discourse, text & metaphor in the representation of landscape.* London: Routledge

Barre, R. (2001). Synthesis of technology foresight. In A. Tübke, K. Ducatel, J. P. Gavigan, & P. Moncada-Paternò-Castello (Eds.), *Strategic policy intelligence: Current trends, the state of play and perspectives* (IPTS-ESTO Report EUR20137EN). Seville: European Commission, Joint Research Centre.

Bauman, Z. (2000). *Liquid modernity.* Cambridge: Polity Press.

Bauman, Z. (2008). *Does ethics have a chance in a world of consumers?* Cambridge, MA: Harvard University Press.

Beck, U. (1992). *The risk society: Towards a new modernity.* London: Sage.

Bertman, S. (2000). *Cultural amnesia: America's future and the crisis of memory.* Westport, CT: Greenwood.

Brown, S. L., & Eisenhardt, K. M. (1997). The art of continuous change: Linking complexity theory and time-paced evolution in relentlessly shifting organizations. *Administrative Science Quarterly*, 42, 1–34.

Bryson J. (2000). Spreading the message: Management consultants and the shaping of economic geographies in space and time. In J. Bryson, P. Daniels, N. Henry, & J. Pollard (Eds.), *Knowledge, space, economy* (pp. 157–175). London: Routledge.

Chapman, A. (2007). *Democratizing technology: Risk, responsibility and the regulation of chemicals.* London: Earthscan.

Cicmil, S., & Hodgson, D. (2006). Making projects critical: An introduction. In D. Hodgson & S. Cicmil (Eds.), *Making projects critical* (pp. 1–28). New York: Palgrave MacMillan.

Cicmil, S., Hodgson, D., Lindgren, M., & Packendorff, J. (2009). Project management behind the facade. *Ephemera: Theory and Politics in Organization*, 9(2), 78–92.

Cunliffe, A. L. (2009). The philosopher leader: On relationalism, ethics and reflexivity? A critical perspective to teaching Leadership. *Management Learning*, 40(1), 87–101.

Cunliffe, A. L., & Easterby-Smith, M. (2004). From reflection to practical reflexivity: Experiential learning as lived experience. In M. Reynolds & R. Vince (Eds.), *Organizing reflection*. Aldershot: Ashgate.

Dauvergne P. (2008).*The Shadows of Consumption: Consequences for the Global Environment*. Cambridge (MA): MIT Press

De Ronde, M., & Miller, A. (2005). Publish or Perish Bane or Boon of Academic Life? Journal of Management Inquiry, 14(4), 321–329.

Dodgson, M. (1991). *The management of technological learning: Lessons from a biotechnology company*. Berlin: Walter de Gruyter.

Dodgson, M. (1993). Learning, trust, and technological collaboration. *Human Relations*, 46, 77–95.

Dodgson, M. (1996). Learning, trust and inter-firm technological linkages. In R. Coombs, A. Richards, P. P. Saviotti, & V. Walsh (Eds.), *Technological collaboration: The dynamics of cooperation in industrial innovation* (pp. 54–75). Cheltenham: Edward Elgar.

Edquist, C. (2010). *The Swedish paradox: Unexploited opportunities!* (Paper no. 5.) Lund: Lund University, Centre for Innovation, Research and Competence in the Learning Economy.

Eisenhardt, K. M., & Tabrizi, B. N. (1995). Accelerating adaptive processes: Product innovation in the global computer industry. *Administrative Science Quarterly*, 40, 84–110.

Engeström, Y. (1987). *Learning by expanding: An activity-theoretical approach to developmental research*. Helsinki: Orienta-Konsultit.

Eriksson, E. A., & Weber, K. M. (2007). Adaptive foresight: Navigating the complex landscape of policy strategies. *Technological Forecasting & Social Change*, 75, 462–482.

Feenberg, A. (1999). *Questioning Technology*. London: Routledge

Felt, U., Wynne, B., Callon, M., Gonçalves, M. E., Jasanoff, S., Jepsen, M., et al. (2007). *Taking European knowledge society seriously: Report of the Expert Group on Science and Governance to the Science, Economy and Society Directorate, Direcotrate-General for Research, European Commission*. Brussels: Directorate-General for Research Science, Economy and Society

Florida, R. (2002). *The rise of the creative class: And how it's transforming work, leisure, community and everyday life*. New York: Basic Books.

Flyvbjerg, B. (2004). Phronetic planning research: Theoretical and methodological reflections. *Planning Theory and Practice*, 5(3), 283–306.

Forsythe, D. (1999). It's just a matter of common sense: Ethnography as invisible work. *Computer Supported Cooperative Work: The Journal of Collaborative Computing*, 8, 127–145.

Fougère, M., & Solitander, N. (2010). Governmentality and the creative class: Harnessing Bohemia, diversity and freedom for competitiveness. *International Journal of Management Concepts and Philosophy*, 4(1), 41–59.

Guston, D. H., & Sarewitz, D. (2002). Real-time technology assessment. *Technology in Society*, 24(1–2), 93–109.

Hasu, M. (2000a). Blind men and the elephant: Implementation of a new artifact as an expansive possibility. *Outlines—The Journal of Critical Social Studies*, 2(1), 5–41.

Hasu, M. (2000b). Constructing clinical use: An activity-theoretical perspective on implementing new technology. *Technology Analysis and Strategic Management*, 12(3), 369–382.

Hasu, M. (2001). *Critical transition from developers to users: Activity-theoretical studies of interaction and learning in the innovation process*. University of Helsinki, Center for Activity Theory and Developmental Work Research. Espoo: Otamedia Oy. Retrieved from http://ethesis.helsinki.fi on June 15th, 2011

Hasu, M. (2005). In search of sensitive ethnography of change: Tracing the invisible handoffs from technology developers to users. *Mind, Culture, and Activity—An International Journal*, 12(2), 90–112.

Hasu, M., & Engeström, Y. (2000). Measurement in action: An activity-theoretical perspective on producer-user interaction. *International Journal of Human Computer Studies*, 53, 61–89.

Hoogma, R., Kemp, R., Schot, J., & Truffer, B. (2002). *Experimenting for sustainable transport: The approach of strategic niche Management*. London: EF&N Spon.

Hyysalo, S. (2004). *Uses of innovation: Wristcare in the practices of engineers and elderly*. University of Helsinki, Department of Education. Helsinki: Helsinki University Press.

Jackson T. (2009). *Prosperity without growth? The transition to a sustainable economy*. Retrieved from Sustainable Development Commission http://www.sd-commission.org.uk/data/files/publications/prosperity_without_growth_report.pdf. Accessed Nov.10.2010

John-Steiner, V. (2000). *Creative collaboration*. New York: Oxford University Press.

Kessler, E., & Chakrabarti, A. (1996). Innovation speed: A conceptual model of context, antecedents, and outcomes. *The Academy of Management Review*, 21(4), 1143–1191.

Kuhlmann, S., & Smits, R. (2004). The rise of systemic instruments in innovation policy. *International Journal of Foresight and Innovation Policy*, 1(1/2), 4–32.

Kuhn, T. S. (1996). *The structure of scientific revolutions*. Chicago: University of Chicago Press.

Lee, R. L. M. (2005). Bauman, liquid modernity and dilemmas of development. *Thesis Eleven*, 83, 61–77.

Lindgren, M., & Packendorff, J. (2006). What's new in new forms of organizing? On the construction of gender in project-based work. *Journal of Management Studies*, 43(4), 841–866.

Markman, G., Gianiodisa, P., Phanb, P., & Balkin, D. (2005). Innovation speed: Transferring university technology to market. *Research Policy*, 34(7), 1058–1075.

McElroy, M. W. (2002). Social innovation capital. *Journal of Intellectual Capital*, 3(1), 30–40.

Meriläinen, S., Tienari, J., Thomas, R., & Davies, A. (2008) Hegemonic Academic Practices: Experiences of Publishing from the Periphery. *Organization*, 15(4), 584–597.

Miettinen, R., & Hasu, M. (2002). Articulating user needs in collaborative design: Towards an activity-theoretical approach. *Computer Supported Collaborative Work*, 11(1–2), 129–151.

Perlow, L. A. (1999). The time famine: Toward a sociology of work time. *Administrative Science Quarterly*, 44, 57–81.

Prahalad, C. K., & Ramaswamy, V. (2004). Co-creating unique value with customers, *Strategy & Leadership*, 32(3).

Ritzer, G., & Jurgenson, N. (2009). 'Production, consumption, presumption: The nature of capitalism in the age of the digital "prosumer."' *Journal of Consumer Culture*, 10(1), 13–36.

Rogers, E. M. (1983). *Diffusion of innovations*. New York: Free Press.

Rose, R. (1993). *Lesson-drawing in public policy*. Chatham, NJ: Chatham House.

Saari, E., & Kallio, K. (2010). Developmental impact evaluation for facilitating learning in innovation networks. *American Journal of Evaluation*, 32(2), 227–245.

Schaufeli, W. B., & Bakker, A. (2004). Job demands, job resources, and their relationship with burnout and engagement: A multi-sample study. *Journal of Organizational Behavior*, 3(25), 293–315.

Schomberg, v. R. (2011). The quest for the "right" impacts of science and technology: An outlook towards a framework for responsible research and innovation. In M. Dusseldorp & R. Beecroft (Eds.), *Technikfolgen abschätzen lehren. Bildungspotenziale transdisziplinärer Methoden.* Wiesbaden: VS Verlag.

Schot, J. (2001). Towards new forms of participatory technology development. *Technological Analysis & Strategic Management*, 13(1), 39–52.

Schot, J. (1992). The policy relevance of the quasi-evolutionary model: the case of stimulating clean technologies. In R. Coombs, P. Saviotti & V. Walsh (Eds), *Technological Change and Company Strategies: Economic and Sociological Perspectives*, London: Academic Press.

Schmidt, J. (2008) Normativity and Innovation: An Approach to Concepts of Innovation from the Perspective of Philosophy of Technology, *School of Public Policy Working Paper Series*, WP#31, Atlanta: Georgia Tech.

Schön, D. (1987) Toward a New Design for Teaching and Learning in the Professions. San Francisco: Jossey-Bass.

Sennett, R. (2006). *The culture of new capitalism.* New Haven, CT: Yale University Press.

Star, S. L. (1991a). Invisible work and silenced dialogues in knowledge representation. In I. Eriksson, B. Kitchenham, & K. Tijdens (Eds.), *Women, work and computerization* (pp. 81–92). Amsterdam: Elsevier.

Star, S. L. (1991b). Power, technology and the phenomenology of conventions: On being allergic to onions. In J. Law (Ed.), *Sociological review monograph: Vol. 38. A sociology of monsters: Essays on power, technology and domination* (pp. 26–56). London: Routledge.

Star, S. L., & Strauss, A. (1999). Layers of silence, arenas of voice: The ecology of visible and invisible work. *Computer Supported Cooperative Work: The Journal of Collaborative Computing,* 8, 9–30.

Sterling, S. (2004). Higher education, sustainability and the role of systemic learning. In P. B. Corcoran & A. E. J. Wals (Eds.), *Higher education and the challenge for sustainability: Problematics, promise and practice.* Hingham, MA: Kluwer.

Stirling, A. (2005). Opening up or closing down: Analysis, participation and power in the social appraisal of technology. In M. Leach, I. Scoones, & B. Wynne (Eds.), *Science and citizens: Globalization and the challenge of engagement* (pp. 218–231). London: Zed.

Stirling, A. (2008). Science, precaution and participation: Towards more 'reflexive governance' for sustainability. *Science, Technology, & Human Values,* 33(2), 262–294. In Festschrift in honour of T. O'Riordan, University of East Anglia.

Sundbo, J., & Fuglsang, L. (Eds.). (2002). *Innovation as strategic reflexivity.* London: Routledge.

Thrift, N. (2005). *Knowing Capitalism.* London: Sage.

Turner, R., Huemann, M., & Keegan, A. (2008). Human resource management in the project-oriented organization: Employee well-being and ethical treatment. *International Journal of Project Management*, 26, 577–585.

Usui, K. (2008). *The Development of Marketing Management The Case of the USA c. 1910–1940.* Aldershot: Ashgate

Vaguer, C., & Van Bastelaer, A. B. (2004). Working overtime. In *Statistics in Focus, Population and Social Conditions.* Dublin: European Foundation for the Improvement of Living and Working Conditions.

van Griethuysen, P. (2010). Why are we growth-addicted? The hard way towards degrowth in the involutionary western development path. *Journal of Cleaner Production,* 18(6), 590–595.

von Hippel, E. (1994). 'Sticky Information' and the Locus of Problem Solving: Implications for Innovation. *Management Science*, 40(4),429–439

Varblane, U., Dyker, D., Tamm, D., & von Tunzelmann, N. (2007). Can the national innovation systems of the new EU member-states be improved? *Post-Communist Economies*, 19(4), 399–416.
Virtanen, M., Ferrie, J. E., Gimeno, D., Vahtera, J., Elovainio, M., Singh-Manoux, A., et al. (2009). Long working hours and sleep disturbances: The Whitehall II prospective cohort study. *SLEEP*, 32(6), 737–745.
Virtanen, M., Ferrie, J. E., Singh-Manoux, A., Shipley, M. J., Vahtera, J., Marmot, M. G., et al. (2010). Overtime work and incident coronary heart disease: The Whitehall II prospective cohort study. *European Heart Journal*, 31, 1737–1744.
Virtanen, M., Singh-Manoux, A., Ferrie, J. E., Gimeno, D., Marmot, M. G., Elovainio, M., et al. (2009). Long working hours and cognitive function: The Whitehall II study. *American Journal of Epidemiology*, 169(5), 596–605.
von Hippel, E. (1994). "Sticky information" and the locus of problem solving: Implications for innovation. *Management Science*, 40(4), 429–439.
von Hippel, E. (1998). Economics of product development by users: The impact of "sticky" local information. *Management Science*, 44(5), 629–644.
von Hippel, E., & Tyre, M. (1995). How learning by doing is done: Problem identification in novel process equipment. *Research Policy*, 24, 1–12.
von Hippel, E., & Tyre, M. (1996). The mechanics of learning by doing: Problem discovery during process machine use. *Technology and Culture*, 37(2), 312–329.
Voß, J.-P., Bauknecht, D., & Kemp, R. (Eds.). (2006). *Reflexive governance for sustainable development.* Cheltenham: Edward Elgar.
Weber K. (2006). Foresight and adaptive planning as complementary elements in anticipatory policy-making: a conceptual and methodological approach. In J-P. Voß, D. Bauknecht, R. Kemp (Eds), *Reflexive Governance for Sustainable Development,* Cheltenham: Edward Elgar,
Williams, R., Slack, R., & Stewart, J. (2005). *Social learning in technological innovation: Experimenting with information and communication technologies.* Cheltenham: Edward Elgar.
Woolthuis, R. K., Lankhuizen, M., & Gilsing, V. A. (2005). A system failure framework for innovation policy design. *Technovation*, 25, 609–619.
Zika-Viktorsson, A., Sundström, P., & Engwall, M. (2006). Project overload: An exploratory study of work and management in multi-project settings. *International Journal of Project Management*, 24, 385–394.

6 Innovation and the Global Financial Crisis

Systemic Consequences of Incompetence

Karl-Erik Sveiby

Intense innovation has changed the financial sector dramatically in the last 30 years. Rajan (2005) identifies four main drivers: regulation-deregulation, which generates 'circumventive' product development (Holland, 1975); institutional change caused by deregulation; globalization made possible by information and communication technology (ICT); and general technical change mainly due to ICT. Benefits of innovation are commonly regarded as an open-ended, linear function of innovation volume and intensity: 'faster diffusion means a higher societal return' (Frame & White, 2004, p. 118). Yet we know that practically all innovations also have less beneficial consequences, which often are systemic in nature, and this reduces the net value for users and at least some members of society (see chapter 4 in this book).

The chapter applies the concept of incompetence by Polanyi (1962) and the concept of unintended consequences by Merton (1936) to explore the development of a radical financial innovation, *securitization*. This innovation kept changing the professional context for financial actors unpredictably for three decades, gradually at first, then accelerating as the drivers kicked in one after the other. Competence was gradually created, destroyed, and recreated so fast that the agents never got the time they needed to become fully competent up until the final crash in 2007. Product developers as well as executives and policy makers were always one step behind; one might say that they were constantly becoming 'temporarily incompetent.' The consequences were that highly skilled designers developed products with built-in errors; very experienced banking executives unwittingly made highly risky decisions; and, highly intelligent and careful policy makers made interventions in the markets that generated opposite effects than they had intended.

The case is a chronological analysis from 1980 until 2008, the year the global financial crisis begins. The analysis focuses on a class of financial instruments developed with securitization known under a collective name, *collateralized debt obligations*[1] (CDO). CDOs are different compared to physical products in that their value depends on predictions about future behaviors of U.S. home-owners and corporations; one might say that the *prediction is the product*. This makes them an ideal product to follow when the research interest is in the relation between competence and innovation.

The first generation of securitization was developed in 1970 by U.S. quasi-governmental agencies[2] to enable a new financial instrument: the *pass-through mortgage-backed bond*—the 'ancestor' of the CDO. The purpose of the innovation was to solve the agencies' funding problem by making mortgage loans more attractive for Wall Street banks. Securitization was originally based on three simple principles. A special purpose vehicle (SPV), a trust controlled by the government agency, was set up, and hundreds of thousands of individual mortgages were transferred from the agency to a pool owned by the SPV. The SPV then issued certificates, each representing a pro rata slice, tranche, of the principal and interest, which were sold to investors (Silber, 1975, p. 106; Finnerty & Emery, 2002).

METHOD AND DATA

The method is to start with a case where the unintended consequences are known and then to "go backwards" in the event chain in search of prediction errors in publicly available sources (see figure 6.1). The research question guiding the analysis is: What evidence of actors' prediction errors can be found? The analysis focuses on critical events when four crucial competencies of financial actors are put to the test: How to evaluate credit risk (that the collateral in the CDO defaults), price risk (that the CDO loses value), liquidity risk (that investors are unable to sell the CDO), and model risk (that the model for designing the CDO does not mirror reality).

Quantitative data[3] from Thomson Banker One and Freepatentsonline are used to illustrate annual innovation volume measured as number of new product or service launches and patent applications. Identified changes in CMO/CDO characteristics are classified as *new product development (NPD)*. Minor changes to characteristics are collectively called *new product offerings*. The CMO/CDO varieties created with securitization are not classified in further detail. Qualitative data are of two kinds: The professional context (see figure 6.2) is operationalized as the six editions of a classic work covering most of the time period, *Handbook of Mortgage-Backed Securities* (Fabozzi, 1985–2006), comprising in total 246 chapters. Journal articles in the *Wall Street Journal* and *Financial Times* going back to 1982 and 1986, respectively, are sources for indications of experts' prediction errors and their consequences. Academic research has several roles: as source of time-dependent facts, as evidence of the professional context in the case narrative, and in the discussion. A glossary with financial terms is found at the end of this chapter as appendix 6.1.

THE RELATIONSHIP BETWEEN INNOVATION, COMPETENCE, AND UNINTENDED CONSEQUENCES

Competence in working life is, following Polanyi (1962), defined as an individual's capacity to act successfully in a professional context. Competence is thus not a property, but a relation between individual actors and professional context; a social system of rules. In finance, professional competence can be seen as a combination of practical skills in the tools of the profession and rule-based problem solving, where some of the rules (heuristics) are explicit in the form of economic/financial theory. Professionals or employees are regarded by their professional peers and colleagues as competent because of the result of their performance. A professional learns the rules gradually through education, learning by doing and reflection.

Competence includes an ability to foresee the consequences of one's actions within the professional context: to make predictions, which turn out to be correct over a period of time. Polanyi (1962) draws a distinction between two kinds of errors, namely professional predictions, which turn out to be mistaken, and unprofessional predictions, which are not only false but *incompetent* (p. 116). The theory implies that a person can become incompetent when a major shift occurs in the professional context, such as when a radical innovation is introduced. The introduction of securitization in the financial industry was such a major technology shift.

In Schumpeterian economics it is generally assumed that competencies, which become obsolete in the process of 'creative destruction,' are replaced by new competencies appropriate for the new context. Innovation thus both destroys and enhances firm level competences (Tushman & Anderson, 1986). However, competence in a context changing new technology has to be learned through learning by doing (Rosenberg, 1976); a process that requires time. Hence, for a while, those learning the new context will not yet be competent; whereas, simultaneously, those competent in the old context are becoming less competent. Both will be making errors, such as predictions that turn out to be false. As the case will show, this is what happened in the financial markets when securitization was introduced. Otherwise competent actors unwittingly made prediction errors, which caused unintended and undesirable consequences (Merton, 1936). The proposed theory derived from extant theory is summarized in figure 6.1.

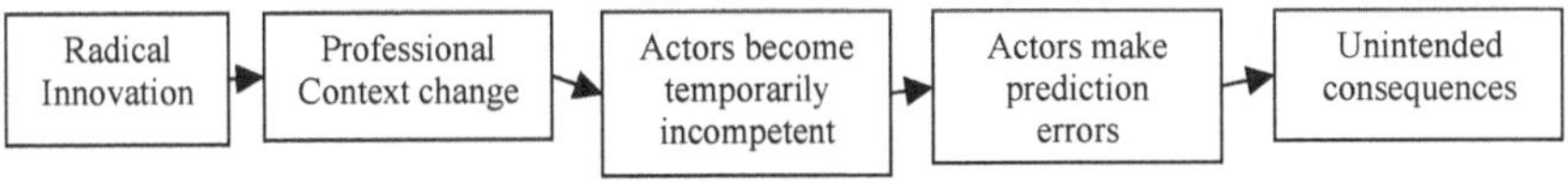

Figure 6.1 Innovation and competence destruction—proposed chain of events.
Sources: Merton, 1936; Polanyi, 1962; Schumpeter, 1939; Tushman & Anderson, 1986.

Financial instruments can be seen as action rules made explicit (Polanyi, 1962); competence formalized in legal contracts and codified in computer models and processes. They are generally developed in close collaboration between customer and provider (Tufano, 2003). Following the conceptualization of Gallouj and Weinstein (1997) and Gallouj and Savona (2009), the CMO/CDO is here seen as a set of material and immaterial technical characteristics created in interactions between the competences of actors in the professional context (see figure 6.1). Together they make up the final characteristics of the CMO/CDO. Changes in the characteristics define several types of service innovation—one of them is *radical innovation*, defined as creation of a completely new set of competences and technical and service characteristics. An example is the Cash Management Account invented by Merrill Lynch, the first financial process to be granted patent in 1983 (Rossignoli & Arnaboldi, 2009). Less recognized, but ultimately of much higher importance for the financial industry (Cobas, Mote, & Wilcox, 2002), is the financial innovation of securitization, a both competence-enhancing and competence-destroying (Tushman & Anderson, 1986) 'technology,' analyzed more closely in this chapter.

THE CASE

The 1980s: The CMO and the CDO Are Developed

By 1980, the pass-through bond is a well-established product and easy to understand for investors, but a vexing problem is how to value the new security. U.S. home owners are free to repay the whole or part of their mortgage loan at any time. Hence, when interest rates go down, they tend to repay earlier, forcing the investors to re-invest the received cash at a lower rate; when rates go up, they tend to repay later, which prevents the investor from re-investing the cash flow at higher rates. The pass-through bond amplifies this *prepayment risk*.

The second generation instrument developed with securitization is the collateralized mortgage obligation (CMO) developed by FHLMC and Bankers Trust (Morris, 2008, p. 39) in 1983. It partly solves the prepayment problem by adding new technical characteristics: The cash flows from principal and interest repayments in the pool are *unbundled* and redistributed so that the highest ranked tranches receive predetermined payouts and are isolated from the risk. However, this means that the cash flow of lower ranked tranches becomes more erratic (i.e., they have to take more risk). The most exposed among these is the 'Z-bond'. It does not receive any cash flow at all until all other tranches have fulfilled their conditions. The CMO looks like the perfect financial product to fill the void in home owner funding created by the savings and loan crisis (Mayer, 1997, pp. 28–31). Hence, government rapidly follows up with new legislation[4] in 1986 to smooth out tax issues (Henderson & Scott, 1988, p. 38).

The general securitization technology is now in place, and the professional context (shaded area) with the main actors looks like figure 6.2. The ultimate purpose of the complex structure is to serve the Borrowers and Other debtors with loan capital; they mortgage their homes and contribute data about themselves to the six actors in The Professional Context. All are dependent on each others' competencies and data, and they are also to a large degree linked by legal contracts. The arrows between them in figure 6.2 refer to competence/data dependence. For example: from Borrowers and Other Debtors to Agent (data about themselves); from Agent to SPV (mortgage contracts, data about mortgagees, management capacity); from SPV to Rating Agencies, Pool Insurers, and Bondholders (information about the mortgages, market maker services); from Rating Agencies to Bondholders (information about ratings). Legal and accounting advice are needed by everyone.

The first actors begin to form the professional context of securitization. They are young and inexperienced; an informal poll conducted at a mortgage-backed securities conference in 1984 notes that less than 10% of the attendees have more than three years of experience (Fabozzi, 1985, p. xiii).

Most actors in the financial industry, new as well as old, have to learn the new competencies required: how to design the issuing legal entity (the SPV), how to originate and transfer the original mortgages, how to design

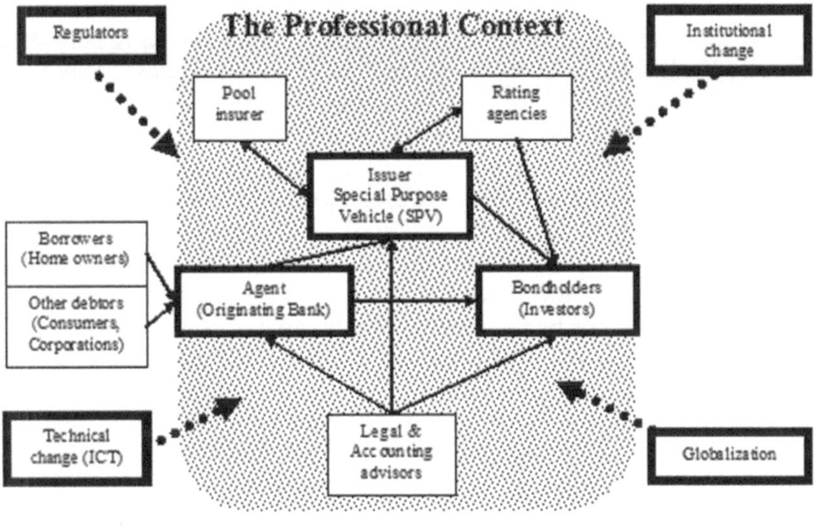

Figure 6.2 Securitization technology and its professional context with four exogenous drivers.

the overall security package, how to ensure that the models mirrored reality (the model risk), how to ensure interest indemnity, how to isolate the bank from bankruptcy of the SPV, how to price the new security, including the pre-payment problem (the price risk), how to calculate and extract profit, and so on (list amended from Henderson & Scott, 1988, pp. 52–53).

The rating agencies have to learn how to assess the credit risk of a pool of diverse mortgages, and their ratings become the only credit risk assessment tool for investors. A major benefit for the investors is that the collateral of the CMOs is guaranteed by the government agencies and that rating institutes also issue ratings. The credit risk is hence considered so low that both investors and the Wall Street banks begin to disregard it. More important for them are the new competence requirements: to evaluate price risk and liquidity risk.

The regulators must learn how to assess the systemic implications of the new financial instruments. This is becoming quite a challenge, because securitization is beginning to cause *institutional change*; the financial industry is breaking up into subsectors. The professional context of 1988 in figure 6.2 is hence becoming even more complex, amended with new institutions offering specialist competences. Mortgage brokers emerge; they solicit and screen loan applicants. Mortgage lenders bid for the loans and hold them until they have collected enough to support a CMO. Investment banks design and market the CMO bonds; they also act as market makers, because, as nonstandardized financial instruments, CMOs are not listed on exchanges. Servicing specialists manage collections and defaults (Morris, 2008, p. 40). Insurance companies and rating agencies begin to explore new markets as CMO pool insurers and CMO raters respectively.

In 1985, the first edition of *The Handbook of Mortgage-Based Securities* with 20 chapters edited by finance professor Frank Fabozzi is published. The range of expertise mirrors the new professional context. Fabozzi is employed by the accounting firm Peat Marwick Mitchell; other authors of the handbook come from investment banks such as Salomon Brothers, Lehman Brothers, First Boston, Morgan Stanley; the stock broker Dean Witter; and the investment advisor Ryan Financial. There is one chapter on evaluating prepayment risk in the handbook. It is written by David Askin, a well-known junk bond expert from Drexel-Burnham-Lambert. He outlines the main problem with most models: they rely on data on historical prepayment patterns collected by the Freddie Mac/Fannie Mae agencies (Fabozzi, 1985, pp. 279–306). Askin is prescient, but he will later become known also in less flattering circumstances.

The 'Originating Banks' in figure 6.2 come under fierce competition from the new intermediaries and profit margins go down. To compensate, they begin to securitize more and more of their mortgages as new CMOs to generate fees instead. A new CMO also removes capital from its balance sheet, whereas the bank still retains influence over its use. This improves a key performance indicator, the return on equity.

By now, the PC brings rapidly growing computing power to the desks of all analysts (Henderson & Scott, 1988, p. 7). A growing cadre of financial analysts begin to apply securitization technology on any type of debt or loan. Fabozzi's handbook (1988) describes four of the innovations; four years later, the third edition devotes no less than 14 chapters of 41 to describe CMO varieties. During the three most active years 1983 to 1985, a new financial instrument is launched every second week—from 1985 increasingly referring to financial products in other currencies than US$; *globalization* adds new innovation opportunities. Because most of these financial products are based on (derived from) the behavior of underlying assets they are called *derivatives*. See figure 6.3.

Investors are happy to invest in tailor-made financial instruments with ratings from AAA downward. Borrowers save money by cutting out the charges made by the banks as middlemen. Rating agencies, pool insurers, and Wall Street Banks see profits and outstanding volume of CMOs surges from around $100 billion in 1987 to around $350 billion by the end of 1990 (Davidson, Sanders, Lan-Ling, & Ching, 2003). Further deregulation allows even the old-style regional banks to join in. They are sitting on a gold mine: not-yet-securitized mortgages.

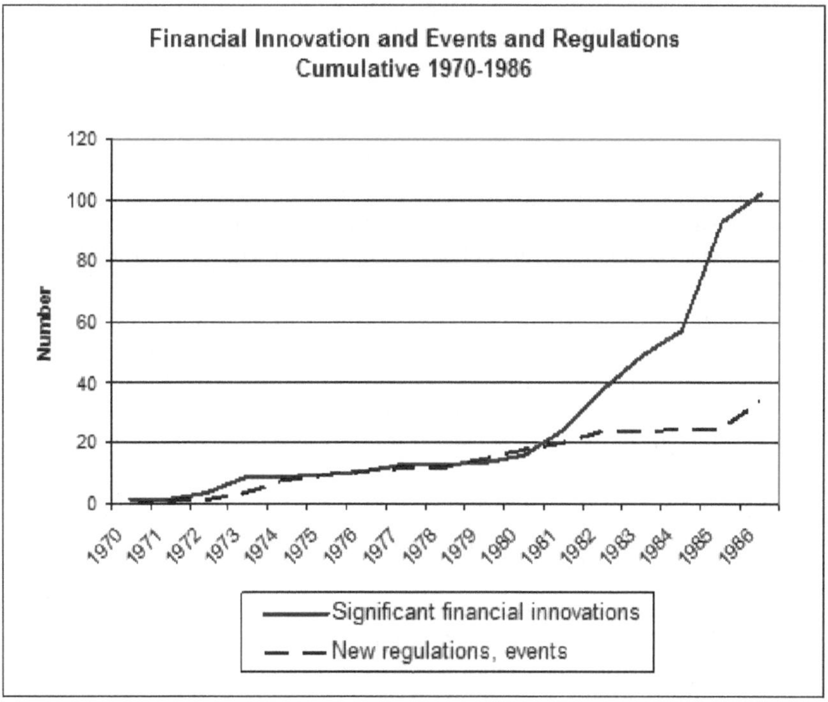

Figure 6.3 Debt innovations 1970–1990.
Sources: Finnerty & Emery, 2002; Matthews, 1994; Miller, 1991.

The U.K. economist Colin Mayer, however, is disturbed by the potential undesirable consequences of the financial innovation wave. In a paper published 1986, he lists among others: the apparent liquidity of the system may prove illusory, inexperience of pricing new instruments may cause mispricing, inadequate perception of risks among managers, and transactions not recorded in accounting statements undermine the soundness of the financial system. His prescient concerns are never tested empirically, neither by his academic colleagues nor by industry experts. But the *price risk* is clearly beginning to be taken seriously by the industry: Eight chapters of 49 in the 1988 edition of Fabozzi's handbook are devoted to prepayment issues. By then securitization, 'this rather ugly word,' has become 'common place in bankers' vocabularies' as Henderson and Scott (1988, p. 1) declare in the first textbook devoted to the topic.

THE 1987 CRASH: POORLY UNDERSTOOD NEW RISKS

In January 1985, a pool insurer goes under, causing losses of some $200 million for Bank of America ('Pacific American Losses,' 1985); and in 1986 and 1987, further major losses[5] follow suit. These are scandals big enough to attract attention from the leading newspapers, but probably only the tip of an iceberg. Culprits are accused, sacked, and sued under headings such as 'fraud,' 'unauthorized activity,' and so on. On October 19, 1987, the U.S. stock market crashes. In the aftermath, multiple unanticipated losses on account of CMOs are uncovered, among them $50 million for First Boston due to errors in valuation. The financial industry is now, for the first time, accused of bringing 'a plethora of new products to market. . . . The scale of the losses suggests a serious lack of understanding about the way such instruments would behave' ('Securitization Takes a More Sober Route,' 1988).

AAA-rated CMO tranches carry no credit risk, but they do contain *price risk*, primarily due to the prepayment problem. Fabozzi (1991, pp. 751ff), contains a study by Jacob and Gallop on previously issued CMOs, noting a 'surprising' result: 'CMO tranches have cash flow patterns that often do not resemble those of more intuitive fixed income securities. The chapter is to be updated with new data under varied authorship in future editions (1995, p. 689; 2001, p. 553; 2006, p. 867), but the conclusion will remain identical despite 15 years of further product development.

Unknown at the time, CMOs also suffers from a completely new *model risk*: The new statistical models produce impressive projections for the behavior of the CMOs, many of which will later be proven to be false. However, the three Fabozzi handbook editions up until 1991 contain no critical analysis on the inherent risk in models. The 1987 crash also highlights that CMOs and other tailor-made securities carry considerable *liquidity risk* when markets turn bad. Faced by more sell-orders than buy-orders, the banks are not able (or simply refuse) to fulfill their role as market makers. 'The enthusiasms of the bull market encouraged illusions about the marketability of securities and distorted perceptions of value. The crash shattered many of these illusions' ('Securitization Takes a More Sober Route,' 1988).

However, banks discover that securitization is a highly effective tool to circumvent the effects of new tougher capital requirements, known as Basle I. The freed up capital is used to create more business of the riskier kind (Franke & Krahnen, 2005).

The 1990s

The main issue in the CMO market is the problem of how to get rid of the 10% lowest classified tranches. None of the top firms will touch them, but the new intermediaries include firms such as Hart Securities Inc., which offer a solution: sell them to unsophisticated clients ('Derivatives Dregs,' 1994). Their high nominal returns also entice an entirely new breed of investors—the hedge funds, which based on projections by statistical models, take highly leveraged bets. Primary among them is a fund run by David Askin, who invests $600 million, mostly borrowed from the CMO-issuing banks, in a highly leveraged portfolio.

By 1994 about $1,000 billion worth of CMOs have been issued, and it is 'the most successful new security product of all time,' as Amerman puts it the first textbook devoted entirely to CMOs (1996, p. 3). By now Bankers Trust and J P Morgan obtain more than half their operating revenues from sources other than lending money. Also, the U.S. government is happy: An estimate made in the mid 1990s concludes that it has saved U.S. home owners $17 billion a year.[6]

Competence requirements are shifting too. An indication is the basic J P Morgan introduction program for new recruits, which no longer includes so much as a day's work on credit assessment (Mayer, 1997, pp. 27–28).

However, hidden from public view, securitization finds also less desirable applications. Around this time, Enron begins building its expansion into gas futures with securitized contracts and SPVs located in offshore tax havens. This eventually becomes the downfall of not only Enron, but also of its advisor, the audit firm Arthur Anderson (Bratton, 2002; Healy & Palepu, 2003).

THE 1994 CRASH

In the spring of 1994, bond markets crash. The triggers are two small increases in short-term interest rates by the Federal Reserve amounting to 0.5%, for the first time in five years. Interest rates soar and bond prices plunge. U.S. President Bill Clinton's chief economic advisor, Robert Rubin, has to admit that he does not know why economic fundamentals do not provide the answer, only that the market is functioning differently than it had in the past ('Stormy Spring,' 1994).

Investors are chocked by the high unknown *price risk* inherent in CMOs: 'The most common complaint I hear nowadays [from investors] is, 'I didn't buy this bond!'' ('Mortgage-Backed Mayhem,' 1994). Clearly, competence in valuing the securities is lacking among all actors. Several highly leveraged hedge funds are again caught by *liquidity risk*. Askin Capital is one of the first to go under. It pulls down with it more venerable firms, such as the 130-year old firm Kidder-Peabody.

The *Wall Street Journal* notes that 'nearly everyone, from the resident White House expert on the markets to the seasoned chairman of the Federal Reserve Board to the managers of the biggest portfolios in the world, seemed confounded' ('Stormy Spring,' 1994). 'This couldn't have happened a decade ago' the *Wall Street Journal* journalist noted. Indeed! Last time U.S. investors were exposed to marketwide *liquidity risk* was only seven years earlier, in 1987. This time, however, the trigger unexpectedly comes from abroad. It turns out that speculators around the world have begun to rely on U.S. treasury markets to hedge their bets. No one has foreseen the huge impact foreign investors could have global on U.S. home market.

The previous quotes indicate that actors are unaware that the professional context has changed considerably. A series of ICT innovations has gradually made the financial world interlinked as never before; in 1994, the traders are making deals based on real-time market data from the financial centers of the whole world. The sudden insight that *globalization*—a term with only positive buzz in financial circles—also meant a world in which an isolated event in a remote corner of the world could have a massive impact across the globe added power to the plunge. 'Herding' has always been an issue in financial markets, but now herding is global and assisted by computers.

The Fabozzi handbook reacts quickly. The 1995 fourth edition devotes no less than 6 chapters (of 39) to prepayment behavior and forecasting and 7 chapters to valuation techniques and risk measurement. However, the pricing problem is still treated as an isolated issue by the experts; the competence to evaluate the *price risk* is clearly still not up to the task, and there is no debate on the problems with statistical modeling and *model risk* on a more fundamental level. The first overview of models will not arrive until 2001. Bykhovsky (2001) writes in the handbook: 'This class of models performed poorly.' They 'dramatically under projected' the volatile behaviors in 1992 and 1993 leading up to the crash. The models were then adjusted and subsequently 'dramatically overprojected' the behaviors in 1996 and 1997; they were adjusted according to the effect that they 'dramatically under projected' the behaviors in 1998 and 1999 (Fabozzi, 2001, pp. 365ff).

The financial industry suffers an estimated $55 billion loss in the 1994 crash, and it will take two more years for the mortgage markets to recover

(Morris, 2008, p. 40). The clean-up also reveals huge losses outside the financial industry, among them pension funds managers and U.S. local councils, which had been attracted by the projected high yields. Merrill Lynch settles a criminal law suit brought by bankrupted Orange County at the cost of $400 million (Shefrin, 2002, p. 204), and both Merrill Lynch and Prudential securities are later fined for inadequate monitoring ('Big Board Disciplines Firms,' 1998).

THE 1997–1998 ASIAN/RUSSIAN CRISES

The mortgage lenders, one of the new classes of intermediary firms created by the securitization boom, are sitting on a stock of $35 billion of mortgage loans that have been originated, but not yet sold as CMO/CDOs, when a financial crisis unexpectedly begins in Asia in July 1997 ('U.S. Banks Set for Big Losses,' 1998). It is followed in August 1998 by a Russian default on their foreign debt.

Interest rates spike, bond prices plummet, and it becomes obvious that the mortgage lenders have underestimated the *price risk*. ContiFinancial is forced to write down $129 million ('ContiFinancial Posts Loss,' 1998), and Mego Mortgage $126 million ('Mego Mortgage Says,' 1998) are two examples of many. They have also been unable to assess *model risk*: Home owners and corporations react by repaying loans much faster than the models predict, and more losses are posted. The 1997–1998 crisis all but wipes out the independent mortgage lenders, but the big banks suffer too. *Model risk* is behind also BankAmerica's $372 million loss in a joint securities trading venture with David Shaw, a former Columbia University computer science professor ('BankAmerica Disaster Arose,' 1998).

Professors are involved also in the decade's most spectacular failure in assessing *model risk*: that of Long Term Capital Management (LCTM), founded in 1994. Boasting two finance professors Myron Scholes and Robert C. Merton (son of social scientist Robert K. Merton) as owners and board members, LTCM's idea is to exploit differences against the efficient market theory wherever they can be spotted. According to LTCM, their models enable them to invest where no one else dares and still make money, and their financial experts predict that only one year in fifty will they lose more than 20% (Lowenstein, 2000, p. 63). Banks queue up to lend them money. In December 1997, when Merton and Scholes received the Nobel Prize in economics for the Black, Merton, Scholes option pricing formula, LTCM's derivative book is worth $1.25 trillion (p. 104) and the firm is leveraged 33 to 1.

However, LTCM's investments are imitated by many other market players globally. LTCM has unwittingly changed the professional context; created a global 'microstructure' (MacKenzie, 2003). LTCM's highly diversified portfolio therefore fails to reduce risk, and all their investments fall in global synchronization in 1998. The 'nature of the world had changed,' says Meriwether, 'and we hadn't recognized it' (cited by MacKenzie, 2003). LTCM is forced to sell, but finds no buyers. The managers had not been able to predict the high *liquidity risk* in their portfolio. Scared by the prospect of a financial system meltdown the Federal Reserve orchestrates a bailout in September 1998. Myron Scholes is unrepentant, however, what had happened was a 'practical impossibility' (cited by Lowenstein, 2000, p. 146).

In 1999, the Gramm-Leach-Bliley Act (1999) removes several of the regulations that had put restrictions on traditional banks since the depression of 1930s. It now becomes possible, among other things, to create bank holding companies, which can be involved in all types of finance, from traditional thrift institutions to investment banking—even insurance.

THE SYNTHETIC CDO AND THE CDS

In 1997, the credit-default swap (CDS) is invented by J P Morgan and the Swiss Bank Corporation (Gibson, 2004). A CDS is the unbundled default risk of a bond. It is sold as a separate product, in essence a derivative of an insurance policy. Credit default swaps can be applied to all kinds of securities and, crucially, can be standardized thanks to ICT. It is but a small step to develop a pool containing credit-default swaps. This innovation, a derivative of derivatives, becomes known as a *synthetic CDO*. Only a year later, new innovations follow: Batches of CDOs and CDSs are pooled into *CDOs squared* and *synthetic CDOs squared*—the first derivative of derivatives of derivatives (Watterson, 2005)!

A particularly attractive feature for the issuers is that synthetic CDOs can be issued without even having to purchase or own the underlying bonds. This is important in Europe where mortgage data are not standardized. Synthetic CDOs therefore become an avenue for European banks to enter the derivatives markets, and this contributes to making the innovation a runaway success. By 2002, synthetic CDOs make up more than 75% percent of issuance the CDO market (Tavakoli, 2008, pp. 6ff), and in 2007, the notional value of outstanding credit default swaps has grown to $62 trillion (Ferguson, 2008, p. 339), a staggering amount, especially considering that the total value of all household and commercial mortgage loans in the U.S. amounts to $14.5 trillion that year (Federal Reserve, 2010).

However, markets are still troubled by price risk and model risk. Increasing interest rates and mortgage prepayments still do not follow historic trends; home owners again react faster than predicted when interest rates increase, and this amplifies downward price swings more than upward swings ('Mortgage Refinancings Go High-Tech Internet,' 1998). Solicited by the banks, many homeowners also begin to borrow more money on the equity in their homes—a new behavior, which has not been covered in the model estimates.

2000–2003, Building up for the Boom

In an article in *Journal of Fixed Income* (2000) the statistician David X. Li proposes what is to become a pivotal invention: a statistical model for estimating future yields of a bond based on life insurance mathematics. The innovation reduces the complexity of calculating prices for pools containing nonstandardized debt, such as corporate bonds, which is exactly what the financial industry is asking for.

However, default rates for corporate bonds in the CDO pools now begin to increase above historic trends (Tavakoli, 2008, p. 10), and in 2002 and 2003, Standard & Poor and Moody downgrade record numbers of such CDOs. The reason: Correlations are not constant as assumed in the models; they vary over time. In 2003, Adelson addresses this problem as *model risk* for the first time in a peer-reviewed article. Markets are unperturbed, however, and innovation accelerates; by 2005, there is 'almost daily innovation in the credit derivative market' (Li, 2006).

Referred to as 'derivatives on steroids,' *CDOs cubed*[7] are introduced in 2005. A year later, they have 'created thousands of new investment assets' (Ray, 2006). By 2008, Merrill Lynch alone had made 384 such CDO deals, in which the underlying assets had undergone 4.79 re-securitizations on

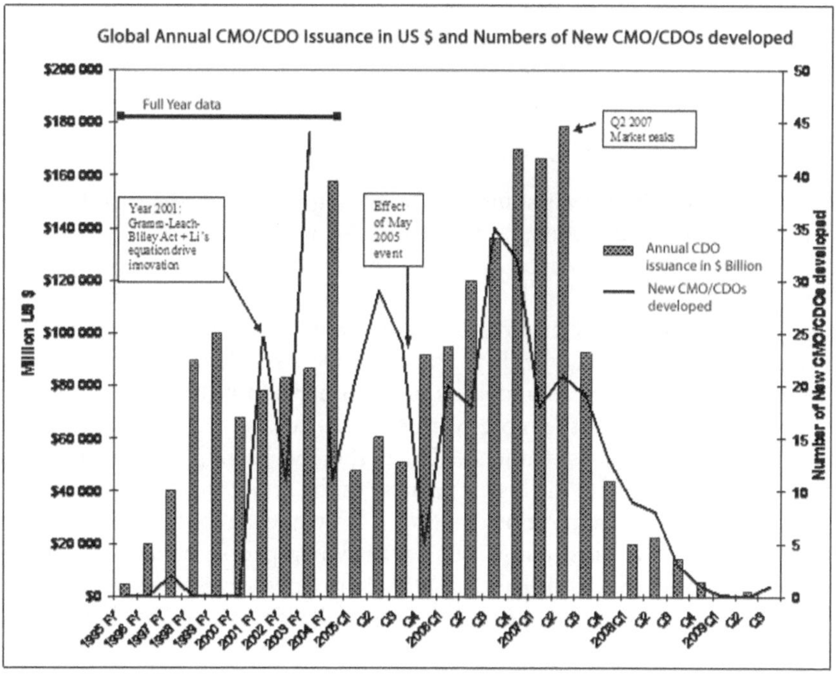

Figure 6.4 Global CMO/CDO innovations and global CDO issuance 1995–2009.

average (Barnett-Hart, 2009). Loans issued to homes owned by the poor, subprime mortgages, a customer segment generally shunned by the big banks, also becomes a new arena for fee growth. CMO pools are filled, aided by the same predatory sales techniques that were applied in the boom leading up to 1994 crash—only now on a much larger scale, and Countrywide Financial, the second largest mortgage lender in the U.S., becomes implicated in a series of lawsuits (Shareholders Foundation, 2010). Employees are pushed to take higher risks in all fields and are even fired if opposing (Kashya, Rajan, & Stein, 2008). Rajan notes in 2005 (p. 339): Overall, 'the incentives to take risk have increased . . . and incentives to herd and move prices away from fundamentals.'

In particular, synthetic CDOs are booming: from a total financial industry risk exposure of less than the equivalent of $400 billion in 2001 to an estimated $2 trillion by the end of 2005, or almost 40% of the whole U.S. corporate-bond market of $4.9 trillion that year, according to J. P. Morgan Chase, one of the most active market players ('Slices of Risk,' 2005).

Fabozzi mirrors the acceleration of innovation in 2006 by expanding the handbook to 54 chapters featuring a new 10-chapter section on 'Alternative mortgage products,' plus 4 chapters on issues related to subprime loans. The book also contains an updated version of Bykhovsky's chapter from 2001 with data up until 2005. It repeats the same criticism about models—not much development seems to have occurred. In total, 69 experts contribute, 23 of them from companies, which will two years later go under or be bailed out by governments: Lehman Brothers, Countrywide, Bear Sterns, Merrill Lynch, United Bank of Switzerland, and Royal Bank of Scotland.

The End Game, 2005–2008

The first serious sign that something is wrong (again) in the CDO market is the 'May 5th 2005 event' (see figure 6.4), when a synthetic CDO pool containing bonds, issued by, among others GM and Ford, causes heavy losses for investors. What had happened? The critical assumption in Li's model is that prices of all credit-default swaps in a synthetic CDO pool move in sync. However, reality proved otherwise. Because only the two carmakers' default risk soared and not the other bonds in the pool, the cost of covering the risk of the more conservative tranches did not rise as much as the cost of the riskiest tranche. Hence, there were no profits from selling those tranches to offset the loss. Li concedes the *model risk* in an interview: 'Actual prices in the market often differ from what the model indicates they should be. . . . The most dangerous part is when people believe everything coming out of it' (Slices of Risk,' 2005). Many investors face serious losses and CDO sales drops in the third quarter of 2005, accompanied by a drastic fall in new product development (see figure 6.4).

In August 2005, central bankers gather for their annual symposium in Jackson Hole, Wyoming, to celebrate the departure of the predominant central banker of the era, U.S. Federal Chairman Alan Greenspan. Raghuram Rajan, chief economist at the International Monetary Fund, destroys the festive mood with his speech titled: *Has Financial Development Made the World Riskier?* (2005). It is regarded as a thinly veiled criticism of U.S. economic policy; and Rajan begins to feel 'like an early Christian who had wandered into a convention of half-starved lions' (Rajan, 2010, p. 3).

Only a couple of weeks later, the bankruptcy of the automotive parts manufacturer Delphi sends another shockwave through the U.S. financial system. Alan Greenspan feels the need to go on record September 27, 2005: 'Recent regulatory reform, coupled with innovative technologies, has stimulated the development of financial products, such as asset-backed securities, collateral loan obligations, and credit default swaps, that facilitate the dispersion of risk' (Greenspan cited by Kothari, 2005). The markets bounce back immediately, and the last quarter of 2005 sees issuance of CMO/CDOs bouncing back (see figure 6.4). This is taken by some as evidence that the rapid growth of hedge funds and of new creative products is making markets more resilient.

Others are less optimistic. 'The events of spring 2005 might not be a true reflection of how these markets would function under stress' , says the annual report of the Bank for International Settlements, BIS, the international organization that coordinates central banks' efforts to ensure financial stability. Citigroup's chairman Chuck Prince: 'When the music stops, in terms of liquidity things will be complicated. But as long as the music is playing, you've got to get up and dance. We're still dancing' ('Bullish Citigroup Is 'Still Dancing,' 2007). Prince is fired, but not because what he had said was untrue.

Prince, like other senior Wall Street bankers, is experienced in riding boom/bust cycles and knows the signs (Ho, 2009). They are bent on 'dancing' until the money flow dries up—and so it does in September 2008 when Lehman Brothers, to the dismay of Wall Street, is not bailed out by the U.S. government. The first ever global liquidity freeze coincides with the other three risks covered in this case: The default risk on subprime loans in times of declining house prices has been underestimated; this type of model risk has not been predicted, so CMO/CDO price risks cannot be calculated.

Aftermath

All the most active players in the CMO/CDO market suffered severe negative consequences, including the two original co-inventors of securitization, Freddie Mac/Fannie Mae, which had to be saved by the U.S. government, while wave after wave of indirect negative consequences rolled through the world economies.

Paul Krugman says: 'Market players seem truly horrified—because they've suddenly realized that they don't understand the complex financial system they created' (Krugman, 2007). The main proponents of the system are like U.S. Federal Chairman Alan Greenspan, 'in a state of shocked disbelief' over the failure of the 'self-interest of lending institutions to protect shareholders' equity' (Clark, & Treanor, 2008), or they blame others like the driving force behind deregulation, Senator Phil Gramm: 'Loose money and politicized mortgages are the real villains' (Gramm: Deregulation and the Financial Panic, 2009). Robert C. Merton is more reflective: 'Is there a structural relation between innovation and crisis? I think there has to be' ('Financial Crisis Caution Urged by Faculty Panel,' 2009).

What happened to the CDO market? After practically dying in 2008, new tougher regulation was put in place and the market revived slightly. Some $2 billion CDOs and CLOs were issued in the first quarter of 2011, a far cry from the $100 to $150 billion quarters of 2006 and 2007. The outstanding stock of asset-backed securities and CDOs therefore began to slowly decrease, down to $2.1 trillion in the first quarter of 2011 (Securities Industry and Financial Markets Association, 2011).

SUMMARY OF THE CASE

By 2008, securitization had become the defining technology of the U.S. financial industry. Traditional banks had shifted from the traditional 'buy and hold the mortgage,' to one of 'trade'; from one where income is generated by the gap in interest between deposit and lending rates to one where income is generated from the difference between buying and selling prices; from one where creditworthiness of the borrower is crucial to one where turnover and trading volume determines the profit (Kregel, 2008). New nonbank financial specialists and intermediaries had evolved into a thriving unregulated U.S.$17 trillion 'shadow banking system,' thereby surpassing the $13 trillion assets of the regulated sector of commercial banks and traditional savings institutions (Adrian & Shin, 2009).

This radical shift was accompanied by severe losses when otherwise competent actors unwittingly made prediction errors due to temporary incompetence. Their prediction errors surfaced during the recurring crises and are summarized in table 6.1.

The case also illustrates several examples of inadequate competence apart from the temporary incompetence of experts. One is the apparent inability of industrial experts to solve inherent problems with the products, such as the prepayment problem, despite numerous efforts. Another is the lack of systemic learning; starting with the 1987 crisis, lack of competence was revealed over and over again during the following 20 years. In the end, it was poor credit risk assessment that triggered the crisis—an ironic twist, because a contributing factor was probably the gradual destruction of the traditional banking core competence when securitization became the main money generator in banking.

Table 6.1 Summary of Prediction Errors and Effect of Innovation Drivers 1970–2008

Competences to assess:	**1970s**	**1980s**	**1990s**	**2000–2005**	**2005->crash**
Model Risk	N/a	N/a	Lacking (1994 and 1997–1998 crashes)	Lacking (mis-appropriation of Li's invention)	Lacking (May 5 event, Subprime debacle)
Price Risk	N/a	Lacking (1987 crash)	Lacking (1994 and 1997–1998 crashes)	Lacking (mis-appropriation of Li's invention)	Lacking (Subprime debacle)
Liquidity Risk	N/a	Lacking (1987 crash)	Lacking (1994 and 1997–1998 crashes)	Lacking (Mounting subprime loans)	Lacking (Subprime debacle)
Credit Risk	High	High	Diminishing	Lacking (Mounting subprime loans)	Lacking (May 5 event, Subprime debacle)
Effect of driver	**1970s**	**1980s**	**1990s**	**2000–2004**	**2005->crash**
Globalization	Not noticeable	Noticed mid-1980s	High (Asian + Russian crises)	Reducedby 9/11 effect	Returning to high
ICT	Beginning	Noticed end 1980s (PCs)	High (Internet)	High (Internet)	High (Internet)
Institutional change	Beginning	High: new fin. firms emerge	High: Turmoil among new fin. firms.	High (New fin. firms disappearing)	Relatively high
Deregulation	Beginning	**High level**	**Continuing**	**Continuing**	**Continuing**
Innovation level	**1970s**	**1980s**	**1990s**	**2000–2004**	**2005-> crash**
	Probably slowly increasing	Very high until 1987, then low	Increasing from low level	Very High	Decreasing before the crash

Table 6.2 Two Waves of Accelerating Financial Innovation Have Occurred since 1970: In the 1980s and the 2000s

Competences to assess:	**1970s**	**1980s**	**1990s**	**2000s**	**Total**
New[a] CMO/CDO offerings	0	59	2	336	397
Repeat CMO/CDO offerings	0	112	7	1,791	1,910
Total CMO/CDO offerings[b]	0	171	9	2,127	2,307
Percent new CMO/CDO *offerings*	14	35%	22%	16%	17%
Significant financial innovations	**15**	**107**	**6**	**n/a**	**127**
Patent applications[c] globally by 20 top CDO banks	0	34	452	1,271	1,772
Patent applications annually		n/a	83	155	

[a]A new CMO/CDO offering is operationalized as a change in at least one of these service characteristics: Lead Manager, Issuer, Type of Security, Description, or Marketplace.
[b]The sharp drop in the number of security codes in the 1990s may partly be due to idiosyncratic classifications in the Thomson database, but it has no implications for the conclusions.
[c]Comparison over time is distorted by changes in patent laws in the U.S. in 1995, when U.S.Patent and Trademark Office began to grant patents for business processes. European patents rules have shifted in the same direction, but are still different. An approximate illustration of innovation activity is the annual average per decade. The 1990s annual average is for the period 1995 to 1999. 2000s annual average is 2000 to 2008.
Sources: Thomson Banker One; Freepatentsonline; author's calculations; Significant financial innovations data: Matthews, 1994; Finnerty & Emery, 2002.

Four decades of financial innovations and CMO/CDO product development are summarized in Table 6.2. Two waves of accelerating innovation can be distinguished: the first in the 1980s, driven by deregulation and ICT. The second wave starts in the late 1990s, when several drivers combine to create an unprecedented acceleration in financial innovation. In relative terms, the most innovative decade was the 1980s with 35% CDO offerings classified as new. The second innovation wave appears a lot less creative; a very high volume of CMO/CDOs was launched, but only 16% could be classified as new[8]; one might say that creativity was halved. The second wave was dominated by customization and combination of innovations originating from the 1980s and 1990s. The innovations became less creative, more distanced from the underlying real assets, and more risky.

DISCUSSION: UNINTENDED AND UNDESIRABLE CONSEQUENCES OF INNOVATION

The various explanations of the financial crisis have by the time of this publishing (2012) come to focus on the U.S. mortgage bubble and the expansion of exotic financial instruments to include subprime loans. Whereas the subprime loan debacle probably was the trigger, this study shows that financial innovations played a crucial role. The following discussion explores the role of lacking competence in this process.

The case largely confirms evolutionary economic theory as a dialectical process between innovation, which generates variety and selection, which alters the relative economic importance of the competing alternatives (Nelson & Winter, 1982; Metcalfe, 1994). We can observe this process as new financial firms emerged and went under continuously due to incompetence in the recurring crises. Securitization forced both new and incumbent actors to create new competence. It also made existing competence less relevant, when the existing industry structure was disrupted (Tushman & Anderson, 1986), such as credit risk assessment skills. The case also shows systemic (Lundvall, 1988) and path-dependent features (Liebowitz & Margolis, 1995; Pierson, 2000a) in that new products were developed through a learning process (Malerba, 1992), which generated new competences among many institutions in a tightly knit professional context. However, the evolutionary dialectic would have led to the extinction of the bank as a species, at least in the Western world.

How could incompetence persist despite three decades of research, intense innovation, and experiential learning? The following discussion applies Robert K. Merton's (1936) framework for 'consequences of purposive action' to try and answer this question. Consequences are both those that are exclusively the direct outcome of the action and those that are the indirect consequences of 'the interplay between the action and the objective [sic] situation.' Merton distinguishes five factors limiting an actor's competence to anticipate direct and indirect consequences of an action for others: lack of foreknowledge, assessment errors, myopia, ideology, and self-defeating/self-fulfilling purpose (see further Gripenberg et al., chapter 1 in this book).

First, in the cause of daily life, humans are limited by *lack of foreknowledge*. It is impossible to foresee all the consequences that may follow an innovation and introduction of new technologies make consequences even less predictable (Redding, 2002). The U.S. agencies could hardly foresee in 1970 that their design to solve a technical issue was to transform the financial industry. Ten years later, it was still only an unlabeled combination of various administrative processes. The shift happened when the financial industry embraced the PC-revolution. In a matter of a few years, the technical processes and the structure were elevated from the humble back office to a generic platform for product development. ICT innovations from then on defined what was possible; they 'reversed' the product development cycle (Barras, 1990). ICT also dramatically accelerated the innovation process itself (Rothwell, 1992). This transformation from unnamed obscurity to 'securitization' explains why banking analysts generally failed to see securitization as an innovation and to predict its importance. The rapid growth of securitization was long interpreted as a decline of the banking sector, because the assets, which were transferred from the banks' balance sheets to the SPVs, disappeared from the financial statistics (Cobas, Mote, & Wilcox, 2002, p. 70).

Beginning in the 1970s institutional change came from deregulation, which gradually allowed banks more freedom and also opened up markets to new financial specialists from the mid-1980s. Globalization reinforced by ICT added further opportunities for innovation from the 1990s. The professional context for securitization hence kept changing unpredictably. Competence was created and destroyed in a tumultuous process, and the actors never got the time they needed to become fully competent up until the final crash in 2007.

The second factor limiting is actors' ability to predict consequences are *assessment errors;* incorrect analysis due to heuristic driven biases, among them the belief that those actions, which have in the past led to the desired outcome, will continue to do so. Many other sources of heuristic biases have later been identified by behavioral economics (e.g., Shefrin, 2002). To the young university graduates, CMO/CDO product development was not very different from an exam in advanced finance; the models were applications of statistical theory they had just learned. There was little appreciation of the issues involved with 'products,' whose value depended on predictions of future behaviors of home owners, the social reality underlying the data.

Whereas any new technology can generate the first type of assessment error, the following is probably unique to the case: CMO/CDO products contain built-in assessment errors. CMO/CDOs are different compared to physical products in that their value depends on predictions about future behaviors of U.S. home-owners and corporations; one might say that the *prediction is the product.* However, the fast growing volume of securitized new products began to influence the behaviors of borrowers on a large scale. People learned to utilize the new services offered, such as refinancing, and new loan products by new financial intermediaries. The Internet added further options. Hence, from around 1993, the historic data underlying the models no longer corresponded to reality. The response was to design new models, only to see them again made obsolete by a changed reality; the designers were always one step behind. What the individual innovators did not perceive was that they, as a group, contributed to a general shift in the financial environment to greater uncertainty, as a study about ratings for bonds issued by banks has shown (Morgan, 2002). Building on data from the 1990s, the econometrician Taleb (2008) concurs: historic data cannot be used for projections, because there is no way to guarantee that the data are not random.

The consequence is that after more than 25 years of development and repeated efforts by the best brains in the industry, many of the CMO/CDO varieties are still not safe to buy. They are 'unsafe at any speed,' to borrow the 1960s U.S. consumer activist Ralph Nader's slogan in his campaign against the U.S. automotive industry's refusal to include safety equipment in American cars. Forty years later, it is normal practice in the financial industry to launch unsafe products on a regular basis.

A third factor limiting our competence is *myopia* (nearsightedness): Actors desiring the beneficial consequences of an action so much, that they are blind to any other consequences. Rather than dispersing risks, securitization became used to add debt to increase short-term profit. Also, a portion of the risk default losses of CDO pools largely remained on the books of the issuing bank (Franke & Krahnen, 2005). As a result, rather than dispersing risk, as the regulating bodies believed, securitization had the opposite effect: to concentrate all the risks in the banking system itself (Adrian & Shin, 2009), only waiting for a trigger. It turned out to be the underestimated credit risk in subprime loan lending, but it could have been any other.

Myopia of experts is illustrated by the six editions of the Fabozzi handbook, which do not contain a single chapter on effects outside the professional context. Practically every chapter in the handbooks deals with the applications of securitization, but only a couple can be said to take more reflective or critical stances. None of the editions published immediately after one of the many financial crises contains a chapter discussing what can be learned from it. At any point in time, a large proportion of the chapters were addressing problems caused by the inadequacy of previous innovations, whereas the fundamental flaws in the products remained unsolved. This suggests that strong path dependence (Liebowitz & Margolis, 1995) and myopia go hand in hand in the case. The global professional context for securitization and the CDO was, due to its dependence on U.S. assets, geographically tightly located on Wall Street in New York. This is known to enhance path dependence, and it was further emphasized in an industry chronically driven by time pressure (Rycroft & Kash, 2002).

Quasi-government agencies and legislators were deeply involved in first co-developing the basic technology in 1970 and 1983, and in later years ironing out legal issues arising from securitization. This suggests that the basic structure was more or less iron clad because innovations breaking the existing legal structure could be subject to legal action—a powerful deterrent. This is confirmed by the data, which show a reduction in creativity (table 6.2). The dominant position of securitization can therefore be an example of path dependent lock-in—an unintended effect of legislation (Greener, 2005; Pierson, 2000a, 2000b).

Ideology, limits competence as, 'there is no consideration of further consequences because of the felt necessity of certain action enjoined by certain fundamental values' (Merton 1936). Economics is not value-free science; on the contrary, it is a highly politicized field. During the period covered by the case, financial policies were dominated by neoclassic economics and two main theories: the efficient market hypothesis (all available information is incorporated in the prices, so they at all times reflect the best possible estimate of the underlying value), and the capital asset pricing model (what we see in the financial markets is the result of rational players' expectations and self-interest played out on efficient markets). The theories became enshrined as fundamental values influencing economic policy both in the U.S. and worldwide. The theories also underlie the statistical/mathematical assumptions of many of derivatives created from the basic CMO/CDO structure. It is beyond the scope of this chapter to discuss validity of the theories. Suffice to say that ideology seems to have played a decisive role in limiting the ability of both scholars and policy makers to foresee the consequences of the theories' practical application.

There is one other fundamental value involved: the pro-innovation bias (Rogers, 1983). This may be one reason why actors believed that the solution to a problem caused by innovation was more innovation. We can detect the same bias in the industry's fascination with statistical theory and ICT. In the late 1980s actors were in awe over the first PC spreadsheets and the first generations of statistical software. When the calculations proved to yield false predictions, more powerful computers and new software were the solutions. Today, 'agent-based economic modeling' for the global economy as a whole is touted by another Nobel Memorial Laureate, Daniel Kahneman ('Model Citizens,' 2010). For this, more powerful computers are required . . .

The fifth factor limiting our competence is a feedback loop, the *self-defeating prophecy* and its opposite, the self-fulfilling prophecy. The response by the financial industry to problems in the first innovation wave from 1983 to 1987 was to create new varieties of the same basic design; new CMO/CDO varieties take up some 20% of all the chapters in the Fabozzi handbooks. Likewise, the reaction of the experts to the price risk problem inherent to the models was incremental innovation: to add more variables, longer data periods, and to combine existing models. New solutions for the valuation problems (price risk) were proposed in every edition up until 2006 (13% of all chapters). With hind-sight we can recognize that such efforts were actually self-defeating. Innovation improved the efficiency of the calculations and accelerated the design of new varieties of the existing design, but it did not address the fundamental flaws in the products. The effect was merely to accelerate the financial industry's descent into its death spiral.

CONCLUSIONS, IMPLICATIONS, AND FUTURE RESEARCH

This study has explored a case of radical innovation, securitization, in the U.S. financial industry. The innovation generated dramatic change in the professional context, which in turn caused otherwise competent actors to make prediction errors. By making a historical analysis of a core line of financial instruments developed with securitization, the CDO, the study has highlighted such temporary incompetence and attempted to show the consequences for industry and society and the relevance for innovation theory in general.

The case illustrates how actors, from novices to laureates, repeatedly made errors in predicting the consequences of their actions. In the early days of the technology, inexperienced novices with limited competence designed primitive early versions of the products containing flaws, which remained despite continuous product development. More experienced bankers and investors, limited by heuristic biases and myopia, launched and invested in products dangerous both for themselves and others over and over again. When reality did not correspond to theory, industry experts tended to disregard the unpleasant information, whereas ideology played a role in limiting the competence of otherwise insightful academic experts and policy makers, who chose to assert their views or to reassure without reflection. Securitization's impact was amplified by drivers, which made contextual change and ensuing uncertainty a more or less permanent feature of the industry.

The consequences of the actors' combined temporary incompetence falsify the common belief that 'faster diffusion means a higher societal return'; faster diffusion can mean a faster journey toward a disaster. The solution to a problem caused by innovation is not necessarily more innovation; innovation may merely multiply the effects of an inherently flawed design.

The trouble for actors is that, at any point in time, all consequences of innovation are possible to know only with hindsight, and then often only with a hindsight bias, something that also this study can be accused of. But the study also shows that there exist options to reduce the risk of errors. One is to be very careful when introducing policies, which may accelerate innovation. Such policies may instead amplify already ongoing contextual change in an industry to disaster level. The case illustrates how ongoing contextual change remains hidden while our view of the new is dimmed by the measuring tools of the old. Hence, as suggested also in chapter 7 of this volume, weak signals, such as the opinions of silenced voices or trends in the fringes, are important. They are easily dismissed, ridiculed, or ignored as change agents push the main stream agenda.

Contributions by independent and contrarian thinkers should be encouraged also in corporate research and development. Very few critical inputs by industry experts addressed the inherent flaws of the CDO varieties in more fundamental ways. Related to this is a recommendation to allow time for breaks to look back, and reflect—a suggestion put forward by in chapter 5 of this volume. Such breaks could have reduced the disturbing tendency to repeat mistakes illustrated in the case. Particularly fruitful moments to learn occur after critical events; the road to global financial meltdown was littered by lost opportunities to do so.

In terms of academic research, a more detailed follow-up of Barras's (1990) reverse cycle product development theory would be valuable. Twenty years after his assertion that financial services are the 'vanguard of the service revolution,' the case demonstrates a major issue with ICT-based service innovation: its systemic nature, which on the one hand is a powerful enabler for further innovation, on the other hand, may sends ripples of unintended consequences through societies on a global scale.

The role of legislation in financial innovation is well worth further study. A hypothesis arising from the case is that a legal structure developed for a specific social benefit may generate a path dependent lock-in effect with systemic impact.

One of the observations while conducting this study was that critical and reflective texts are few and far between in the financial field. It is not surprising that industry insiders have an uncritical bias. However, financial academic research does not perform much better. Could the close liaison between the 'Ivy League' universities and Wall Street firms described by Karen Ho (2009) have anything to do with this? A study focusing specifically on the issue should yield interesting results.

Our limited ability to foresee the consequences of one's actions is a fundamental property of being human. It should hence also be seen as a fundamental property of innovation and in product development. Innovation has such a profound impact on future society that practitioners and change agents, as well as researchers and policy makers, should embrace all possibilities to broaden their abilities to foresee, not only the desired outcomes, but also the unintended and the undesired. Rather than being ignored, they should be acknowledged for what they are: a large, often untapped, resource for improving the net effects of innovation for society.

APPENDIX 6

Glossary of Financial terms

Asset-backed securities	A generic term referring to securitized financial instruments containing any type of assets as collateral
CDO Collateralized debt obligation	A contract where investors are promised by an SPV to receive cash flows in a pre-agreed sequence from a pool of any kind of loans. It is the 3rd generation securitization product.
CDO cubed	A CDO containing CDOs squared.
CDO squared	A CDO containing other CDOs.
CLO Collateralized loan obligation	A CDO containing bank loans.
CMO Collateralized mortgage obligation	CMO is the 2nd generation securitization product; one can call it a CDO which contains only mortgages. Cash flows can be distributed in many different ways.
Collateral	Property or goods used as security for a loan.
Collateralize	To pledge property or goods as security for a loan.
Credit default swaps (CDS)	A form of insurance that protects a lender if a borrower of capital defaults on a loan. When a lender purchases a CDS from an insurance company, the

Continued

	liability of the loan becomes a credit that may be swapped for cash upon the loan defaulting. The difference between a traditional insurance policy and a CDS is that anyone can purchase one, even those who have no direct interest in the loan being repaid.
FHLMC 'Freddie Mac'	Federal Home Loan Mortgage Corporation.
FNMA 'Fannie Mae'	Federal National Mortgage Association.
GNMA 'GinnieMae'	Government National Mortgage Association.
Mortgage	Loan agreement (contract) secured by property, for instance a home.
Mortgage broker	An intermediary who brokers mortgage loans. For instance, a home owner can ask a mortgage broker to find the best mortgage loan deal.
Mortgage-backed security	A term referring to all securitized financial instruments containing mortgages as collateral, an example is the CMO.
Originator (agent)	The entity holding the original collateral in a CDO Mostly a bank.
Pass-through-mortgage-backed bond	The 1st generation securitization product. Cash flows in a mortgage pool are redistributed pro-rata.
Prepayment risk	The risk that home owners repay too early or too late compared to the formula used for calculating the value at the time the investor bought the CDO.
Principal	The original loan amount or the remainder after earlier repayments.
Pro rata	A fixed proportion representing the average of interest, risk and repayments of a pool of mortgages.
Securitization	A technique to collect any kinds of loans in a pool, issue certificates, which represent a slice of the pool, and sell to investors.
SPV Special purpose vehicle	A trust or company set up specifically for the purpose of holding a pool of financial instruments. It is normally controlled by the originator of the pool, such as a bank.
Synthetic CDO	A CDO, where the pool contains credit default swaps.
Tranche	A portion of a pool with loans or other debt, such as mortgages.
Unbundle	To create new financial instruments out of the components of another and sell them separately. For instance, divide the cash flow from a mortgage into two instruments: one that receives all interest payments and one that receives the repayments of the loan.
Z-bond	The last tranche in a CDO, which does not receive any cash flow at all until all other tranches have fulfilled their conditions.

NOTES

1. There is a proliferation of terminology in literature. Collateralized debt obligation, CDO, gradually became the accepted term for ABSs (asset-backed securities) backed by a broad variety of collateral mixes such as: CLO (collateralized loan obligation) only bank loans; CBO (collateralized bond obligation) corporate bonds; CFO, (collateralized funds obligation) hedge fund certificates; etc. A CMO (collateralized mortgage obligation) contains only mortgages. Hurst (2000) dates the term CDO to late 1997.
2. Federal National Mortgage Association (FNMA 'Fannie Mae'), the Federal Home Loan Mortgage Corporation (FHLMC 'Freddie Mac') and the Government National Mortgage Association (GNMA 'Ginnie Mae').
3. All securities with a security code classified as 'collateralized' in Thomson Banker One's database 1985–2009, in total 2,307. Data extracted from Freepatentsonline from filings by the 20 top CMO/CDO-issuing banks: 1,772 U.S. patent applications, European Parliament documents, abstracts of Japan, and World Intellectual Property Organization.
4. In legal terms, most of the CMOs hereafter were designed as 'real estate mortgage investment conduits,' REMICs.
5. Account of how in 1986, a Merrill Lynch senior trader incurs a $250 million trading loss ('Merrill Sacks Trader,'1987) and First Boston loses $100 million ('Too Far Too Fast,' 1987).
6. Cited by Morris (2008, p. 40).
7. 'CDOs Cubed' are Collateralized Debt Obligations with a Russian shell doll structure, containing other CDOs which, in turn, contain other CDOs which, finally, contain investment assets of all types. If the innermost "doll" securities are credit default swaps the instrument will be a derivative of a derivative of a derivative of a derivative.
8. The estimate that 16% CMO/CDO offerings in the 2000s were new products concurs roughly with Tufano's (2002) estimate that 20% of all derivatives at that time were innovations, which lends some level of validity.

REFERENCES

Adelson, M. (2003). CDO and ABS underperformance: A correlation story. *The Journal of Fixed Income, 13, 53–63.*

Adrian, T., & Shin, H. S. (2009). *The shadow banking system: Implications for financial regulation* (Federal Reserve Bank of New York, Staff Report No. 382). Retrieved July 2009, from http://www.newyorkfed.org/research/staff_reports/sr382.pdf

Amerman, D. R. (1996). *Collateralized mortgage obligations.* New York: McGraw-Hill.

BankAmerica disaster arose from a lack of standards. (1998, October 15). *Wall Street Journal.*

Barnett-Hart, A. K. (2009). *The story of the CDO market meltdown: An empirical analysis.* Bachelor of arts thesis, Harvard College, Cambridge, MA. Retrieved December 11, 2010, from http://www.hks.harvard.edu/m-rcbg/students/dunlop/2009-CDOmeltdown.pdf

Barras, R. (1990). Interactive innovation in financial and business services: The vanguard of the service revolution. *Research Policy*, 19, 215–237.

Big board disciplines firms individuals. (1998, October 5). *Wall Street Journal.*

Bratton, W. (2002). Enron and the dark side of shareholder value. *Tulane Law Review,* 76, 1275–1361.

Bullish Citigroup is 'still dancing' to the beat of the buy-out boom. (2007, July 10). *Financial Times*.

Bykhovsky, M. (2001). Overview of recent prepayment behavior and advances in its modeling. In F. J. Fabozzi (Ed.), *Handbook of mortgage-backed securities*. New York: McGraw Hill.

Clark, A., & Treanor, J. (2008, October 23). Greenspan: I was wrong about the economy. Sort of. *The Guardian*.

Cobas, M., Mote, L. R., & Wilcox, J. A. (2002). A history of the future of banking: Predictions and outcomes. In B. E. Gup (Ed.), *Future of banking*. Portsmouth, NH: Greenwood.

ContiFinancial posts loss of $114.3 million cites market upheaval. (1998, November 17). *Wall Street Journal*.

Davidson, A., Sanders, A., Lan-Ling, W., & Ching, A. (2003). *Securitization: Structuring and investment analysis*. New York: Wiley Finance.

Derivatives dregs: Houston firms sold risky 'toxic waste' for Wall Street giants. (1994, December 20). *Wall Street Journal*.

Fabozzi, F. J. (Ed.). (1985). *Handbook of mortgage-backed securities* (1st ed.). Chicago: Probus.

Fabozzi, F. J. (Ed.). (1988). *Handbook of mortgage-backed securities* (2nd ed.). Chicago: Probus.

Fabozzi, F. J. (Ed.). (1991). *Handbook of mortgage-backed securities* (3rd ed.). Chicago: Probus.

Fabozzi, F. J. (Ed.). (1995). *Handbook of mortgage-backed securities* (4th ed.). Chicago: Probus.

Fabozzi, F. J. (Ed.). (2001). *Handbook of mortgage-backed securities* (5th ed.). New York: Mcgraw-Hill.

Fabozzi, F. J. (Ed.). (2006). *Handbook of mortgage-backed securities* (5th ed.). New York: Mcgraw-Hill.

Federal Reserve. (2010). *Credit market debt outstanding*, table L.218. Retrieved December 12, 2010, from http://www.federalreserve.gov/releases/z1/Current/z1r-4.pdf

Ferguson, N. (2008). *The ascent of money: A financial history of the world*. London: Allen Lane.

Financial crisis caution urged by faculty panel. (2009, September 29). Harvard Business School newsletter. Retrieved December 29, 2010, from http://hbswk.hbs.edu/item/6013.html

Finnerty, J. D., & Emery, D. R. (2002, Spring/summer). Corporate securities innovation: An update. *Journal of Applied Finance*, 12, 21–47.

Frame, W. S., & White, L. J. (2004). Empirical studies of financial innovation: Lots of talk, little action? *Journal of Economic Literature*, 42, 116–144.

Franke, G., & Krahnen, J. P. (2005, August 18). Default risk sharing between banks and markets: The contribution of collateralized loan obligations. *Konstanz Universitaet/Goethe-Universität and CEPR*. Retrieved November 19, 2010, from http://www.defaultrisk.com/pp_crdrv_51.htm

Gallouj, F., & Savona, M. (2009). Innovation in services: A review of the debate and a research agenda. *Journal of Evolutionary Economics*, 19, 149–172.

Gallouj, F., & Weinstein, O. (1997). Innovation in services. *Research Policy*, 26, 537–556.

Gibson, M. S. (2004). *Understanding the risk of synthetic CDOs* (Working Paper No. 36, Finance and Economics Discussion Series). Washington, DC: Federal Reserve.

Gramm: Deregulation and the financial panic: Loose money and politicized mortgages are the real villains. (2009, February 20). *Wall Street Journal*.

Gramm-Leach-Bliley Act. (2011, November 10). In *Wikipedia, The Free Encyclopedia*. Retrieved January 21, 2009, from http://en.wikipedia.org/wiki/Gramm-Leach-Bliley_Act

Greener, I. (2005). State of the art: The potential of path dependence in political studies. *Politics*, 25, 62–72.
Healy, P. M., & Palepu, K. G. (2003). The fall of Enron. *The Journal of Economic Perspectives*, 17, 3–26.
Henderson, J., & Scott, J. P. (1988). *Securitization*. Cambridge: Woodhead-Faulkner .
Ho, K. (2009). *Liquidated: An ethnography of Wall Street*. Durham, NC: Duke University Press.
Holland, R. (1975). Speculation on future innovation: Implications for monetary control. In W. S. Silber (Ed.), *Financial innovation*. New York: New York University, Conference on Financial Innovation Proceedings.
Hurst, R. R. (2000). Collateralized debt obligations (CDOs): Identity crisis. *The Securitization Conduit*, 3, 3–4.
Kashya, A. P., Rajan, R., & Stein, J. (2008). *Rethinking capital regulation*. Retrieved November 20, 2011, from http://www.kansascityfed.org/publicat/sympos/2008/KashyapRajanStein.09.15.08.pdf
Kothari, V. (2005). Blog. Retrieved December 12, 2010, from http://credit-deriv.com/crenewsfeb04.htm#delphi_downgrade
Kregel, J. (2008). *Changes in the U.S. financial system and the subprime crisis* (Working Paper No. 530, The Levy Economics Institute Working Paper Collection). Retrieved November 20, 2011, from http://www.levyinstitute.org/pubs/wp_530.pdf
Krugman, P. (2007, December 3). Innovating our way to financial crisis. *New York Times*.
Li, D. X. (2000). On default correlation: A copula function approach. *The Journal of Fixed Income*, 9, 43–54.
Li, D. X. (2006). It's all about credit. *North American Actuarial Journal*, 10, ii-iv.
Liebowitz, S. J., & Margolis, S. E. (1995). Path dependence, lock-in, and history. *Journal of Law, Economics and Organization*, 11, 205–226.
Lowenstein, R. (2000). *When genius failed: The rise and fall of long-term capital management*. New York: Random House.
Lundvall, B.-Å. (1988). Innovation as an interactive process: From user-producer interaction to the National System of Innovation. In G. Dosi (Ed.), *Technical Change and Economic Theory*. London: Pinter.
MacKenzie, D. (2003). Management and the sociology of arbitrage. *Economy and Society*, 32, 349–380.
Malerba, F. (1992). Learning by firms and incremental change. *Economic Journal*, 102, 845–859.
Matthews, J. O. (1994). *Struggle and survival on Wall Street*. Oxford: Oxford University Press.
Mayer, C. (1986). The assessment: Financial innovation: Curse or blessing? *Oxford Review of Economic Policy*, 2(4), i–xix.
Mayer, M. (1997). *The bankers: The next generation*. New York: Penguin.
Mego Mortgage says it expects to report loss for fiscal year. (1998, November 30). *Wall Street Journal*.
Merrill sacks trader after loss of dollar $250M. (1987, April 30). *Wall Street Journal*.
Merton, R. K. (1936). The unanticipated consequences of purposive social action. *American Sociological Review*, 1, 894–904.
Metcalfe, J. S. (1994). Evolutionary economics and technology policy. *The Economic Journal*, 104, 931–944.
Miller, M. H. (1991). *Financial innovations and market volatility*. Cambridge: Blackwell.

Model citizens: Building SimEarth. (2010, November 4). *New Scientist.*

Morgan, D. P. (2002). Rating banks: Risk and uncertainty in an opaque industry. *The American Economic Review,* 92, 874–888.

Morris, C. R. (2008). *The trillion dollar meltdown.* Oxford UK: Perseus/Public Affairs Ltd.

Mortgage-backed mayhem. (1994, July). *Institutional Investor.*

Mortgage refinancings go high-tech internet other factors put pace near record. (1998, August 24). *Wall Street Journal.*

Nelson, R. R., & Winter S. G. (1982). *An evolutionary theory of economic change.* Cambridge, MA: Harvard College.

Pacific American losses may be $200M. (1985, January 31). *Financial Times.*

Pierson, P. (2000a). Increasing returns, path dependence, and the study of politics. *American Political Science Review,* 94, 251–267.

Pierson, P. (2000b). The limits of design: Explaining institutional origins and change. *Governance,* 13, 475–500.

Polanyi, M. (1962). *Personal knowledge: Towards a post-critical philosophy.* Chicago: University of Chicago Press.

Rajan, R. (2005). *Has financial development made the world riskier?* (NBER Working Papers Series No. 11728, Cambridge: National Bureau of Economic Research). Retrieved November 20, 2011, from http://papers.nber.org/papers/w11728

Rajan, R. (2010). *Fault lines: How hidden fractures still threaten the economy.* Princeton, NJ: Princeton University Press.

Ray, R. (2006). CDOs cubed: The first-ever triple derivative. *Derivatives Use, Trading Regulation,* 12, 183–189.

Redding, S. (2002). Path dependence, endogenous innovation and growth. *International Economic Review,* 43, 1215–1248.

Rogers, E. (1983). *Diffusion of innovations* (3rd ed.). New York: Free Press.

Rosenberg, N. (1976). On technological expectations. *The Economic Journal,* 86, 523–535.

Rossignoli, B., & Arnaboldi, F. (2009). Financial innovation: Theoretical issues and empirical evidence in Italy and in the UK. *International Review of Economics,* 56, 275–301.

Rothwell, R. (1992). Successful industrial innovation: Critical factors for the (1990) s. *R&D Management,* 22, 221–239.

Rycroft, R. W., & Kash, D. E. (2002). Path dependence in the innovation of complex technologies. *Technology Analysis & Strategic Management,* 14, 21–35.

Securities Industry and Financial Markets Association. (2011). www.sifma.org

Securitization takes a more sober route. (1988, April 7). *Financial Times.*

Shareholders Foundation. (2010). Retrieved November 22, 2010, from http://shareholdersfoundation.com/case/countrywide-mortgage-backed-securities-certificate-investor-lawsuit

Shefrin, H. (2002). *Beyond greed and fear: Understanding behavioral finance and the psychology of investing.* Oxford: Oxford University Press.

Schumpeter, J. (1939). *Business cycles: A theoretical, historical and statistical analysis of the capitalist process* (1st ed.). New York: McGraw-Hill.

Silber, W. S. (1975). Towards a theory of financial innovation. In *Financial innovation.* New York: New York University, Conference on Financial Innovation Proceedings, D. C. Heath.

Slices of risk: How a formula ignited market that burned some big investors. (2005, September 12). *Wall Street Journal.*

Stormy spring: Three-month tumult in bonds lays bare new financial forces. (1994, May 20). *Wall Street Journal.*

Taleb, N. N. (2008). *Fooled by randomness: The hidden role of chance in life and in the markets.* New York: Random House.

Tavakoli, J. (2008). *Structured finance and collateralized debt obligations: New developments in cash and synthetic securitization*. New York: Wiley Finance.
Too far too fast. (1987, July14). *Financial Times*.
Tufano, P. (2003). Financial innovation. In G. M. Constantinides, M. Harris, & R. M. Stulz (Eds.), *The handbook of the economics of finance*. Amsterdam: Elsevier.
Tushman, M. L., & Anderson, P. (1986). Technological discontinuities and organizational environments. *Administrative Science Quarterly*, 31, 439–446.
U.S. banks set for big losses on mortgage loans. (1998 October 16). *Financial Times*.
Watterson, P. N., Jr. (2005). The evolution of the CDO squared. *The Journal of Structured Finance*, 11, 6–12.

7 Weak Signals for Opting Out of the Innovation Race

Karl-Heinz Leitner

The need for innovation has become a necessity in modern economy and society, and it is commonly argued that ultimately individuals, firms, and nations have to innovate. Although, undoubtedly innovation solves everyday life problems, provides enjoyable products, creates wealth, and drives economic development, the question of the intensity and direction of innovation becomes increasingly relevant when innovation becomes all-encompassing. In this chapter, we will examine whether there are signs that some actors of recent economy and society aim to slow down the pace of innovation and what strategies companies chose to decelerate the innovation process and compete successfully at the same time without adopting common innovation strategies.

This chapter originates in findings from an international research project that investigates different organizational patterns of innovation processes and their future perspectives.[1] In the course of this work, various sources (such as academic literature, magazines, newspapers, and Internet sites) have been analyzed to identify how private and public actors create new ways to organize innovation. Based on a collection of identified international practice examples from industry and society, a number of innovation visions, that is, visions for novel innovation processes, were developed that describe how innovation may be organized in a new way in very different fields in the future. The identified models are mostly varying patterns of a very generic trend toward a networked and open form of organizing innovation involving different actors. In the academic literature, such forms of innovation have been labeled as open innovation (Chesbrough, 2003), user innovation (von Hippel, 2005), community-based innovation (Tuomi, 2002), or crowdsourcing (Howe, 2008).

Even though most of the collected signals identified point to an acceleration of innovation processes, there is also some evidence that organizations or individuals intentionally chose strategies to escape from the ongoing innovation race. We will describe different innovation strategies and organization models that are likely to become more important in the future and will examine under which conditions these new innovation models may at the same time slow down the innovation pace. Thereby, we will also compare our findings with recent innovation literature that supports some presented arguments for rethinking and redesigning traditional innovation strategies. We will pick up and discuss a number of innovation visions from the previously mentioned project and examine some of the consequences that have to be considered when opting out of the innovation race.

One strategy we observed during our project is, for instance, that companies do not change their products deliberately to give them a more traditional touch to address the demand for vintage products. Companies may also hide their efforts at making a product look like it is unchanged, focusing more strongly on more invisible process innovation. So called pseudo-innovation strategies that are just superficial adaptations of the same product design are further ways companies can pursue to adapt to the innovation race. A rather traditional strategy to cope with the increasing innovation speed and shortened product life cycles is to apply patents that are then not commercialized. Furthermore, innovation studies have dealt with the phenomenon that firms often do not innovate because they fear that existing products might get cannibalized. Outsourcing innovation activities to specific organizations or user communities is an alternative strategy that allows companies to keep a stable culture while transferring the often associated 'creative turbulence and uncertainty' to specific subindustries, companies, or groups of society. As a consequence, a firm's corporate culture is also changing, with less pressure than other groups of society to innovate constantly. In the context of societal aims to achieve a more sustainable development, we also can observe that some actor groups and academics advocate to slow down the innovation process as such, to be more reflective in technological development and diffusion and foster only very specific types of innovation heading towards sustainable development.

The chapter is structured as follows: At the beginning we will give an overview of the foresight project on innovation futures. Based on these findings, we will present and discuss in more detail different ways to opt out of the innovation race. This is first to deliberately follow a noninnovation strategy, and second to choose an imitation strategy, which may be facilitated by the emerging dominance of China and India as new innovation hubs. Third, we can expect firms, customers, and citizens with their aim to achieve true sustainable solutions will reduce innovation output. Fourth and finally, the trend toward more open and distributed patterns of innovation may slow down innovation in specific contexts. We will then summarize our arguments to conclude the chapter.

NEW INNOVATION PATTERNS: EVIDENCE FROM A FORESIGHT PROJECT

The Innovation Futures (INFU) project deals with the emergence of new innovation patterns, such as open innovation, user innovation, community-based innovation, and design innovation. While these and similar innovation patterns have been discussed intensively in recent years, there is little systematic exploration about possible future innovation landscapes and long-term implications for economy and society.

INFU is a foresight project. Foresight became popular in the 1980s and combines different instruments and methods to surmount the traditional planning and forecasting approach. Foresight activities emphasize the systematic exploration of future dynamics and the importance of interaction between actors from different constituencies in the respective innovation system. Foresight is a method of prospective analysis and informed decision making that includes long- to midterm considerations of likely, possible, or even just thinkable futures (Miles, 2008). Foresight hence does not want to prescribe the future but aims to initiate a critical debate about possible future developments. Foresight exercises have been organized by organizations, interest groups, and policy makers on the regional, national, or supranational level and have become increasingly popular in the last three decades. A recent monitoring study, for instance, has collected 6,000 foresight projects that have been conducted in Europe since 1980 (European Commission, 2008). A wide range of methods already exists that are often combined in different ways for a specific foresight project. Most commonly employed techniques are expert panels, roadmapping, Delphi studies, scenario writing, backcasting, literature reviews, bibliometric searches, modeling, and simulation. The INFU project combines the methods scanning weak signals, organizing expert panels, conducting interviews, and constructing scenarios to construct plausible, relevant, long-term scenarios of future innovation landscapes.

INFU deals with the future of innovation that first of all requires a definition of innovation. The project team has a broad understanding of innovation as encompassing the economic, social, and public domains. The research is interested in how the process of the creation, development, and introduction of innovations is changing and concentrates on the process of 'innovating innovation.' An innovation pattern is defined as the underlying principle on how the innovation process is organized, which also includes new perceptions about what is seen as innovation, the involvement of new actors in the innovation process, and the generation of new interpretations of the novel and the new in society. The process of the emergence of new innovation patterns and change of the innovation landscape is not necessarily a radical one, however. By 'new innovation patterns,' the project team hence means novel emerging concepts, ideas, and strategies of how innovation is organized, but also well-known trends, such as open source software development as alternative way to innovate, which are important in specific industries or areas, but may have a larger impact on or potential for other areas in the future (Jégou et al., 2009). Although open source software design, for instance, is not new, its design principles may be applied for the first time in new areas or industries.

Based on an analysis of various sources (such as academic literature, the Internet, blogs, newspapers, and magazines), 68 'weak signals' for emerging innovation patterns have been identified in the first phase of the project (Jégou et al., 2009). A weak signal is defined as a hint of the potential for change with a possible disruptive impact, which is already visible but has not yet reached the mainstream. In our context, a weak signal hence indicates a change in an innovation pattern with a high potential for disruptive impact, which is not established as a common way of innovating in a certain sector.

The identified signals for change have revealed that many of the observed practices can be referred to as concepts or strategies that have already been described in the academic literature, but often combine elements or strategies in a new manner or realize strategies for an entirely new application field and thus expand the thinking about possible innovation futures. In this context, concepts such as open innovation (Chesbrough, 2003), user innovation (von Hippel, 2005), value innovation (Kim & Mauborgne, 1999), soft innovation (Stoneman, 2007), crowdsourcing (Howe, 2008), and social innovation (Mumford, 2002) can be mentioned that have been defined and debated intensively in the academic literature.

Based on the scanning of weak signals regarding the change of innovation patterns and reviewing the recent academic literature dealing with new forms of innovation, the INFU research team has generated and visualized 20 innovation visions. These visions were formulated in a creative way by amplifying and combining some of the signals to develop coherent and sometimes provocative pictures of possible future innovation practices. Thereby, the team transferred an idea already applied to other sectors or generalized a signal considered to become a mainstream innovation practice.[2] To provoke discussion some visions were brought to an extreme. In addition, the team conducted interviews with 25 experts from industry and academia and organized an online-survey to discuss and assess the innovation visions.

In the following, some innovation visions which are of particular interest for our topic are briefly presented:

By the '*open source innovation model,*' the team envisaged that open source software design such as the development of the LINUX computer software (which competes successfully with Microsoft) has become a significant trend. This innovation vision assumes that open source development is no longer limited to software design but becomes an all-encompassing innovation pattern. Many products and services are provided by a large number of people who contribute different bits and pieces to various technological and social innovation projects.

With '*automated innovation,*' it is assumed that a number of new techniques, such as semantic web analyses, allows for automating parts of the innovation process, ranging from idea generation via design to testing. In this model, sophisticated semantic web filters track changes in consumer preferences and new ideas in real time and automatically extract innovations with outstanding market potential.

The vision *'relocated innovation'* envisages that the West adopts the role of a follower, whereas Asian countries become innovation leaders. Western companies wishfully have to face products primarily designed for a different cultural context and imitate and copy products from Asia. Creative people might even migrate to the new innovation hot spots in Asia and send back their salaries home to the U.S. and Europe. The current tendencies of 'globalization of wisdom' would hence be limited by specialized regional innovation clusters. In addition, Western nations would lose wealth, whereas people in the Middle East and Asia would benefit.

'Innovation marketplace' is described as an innovation pattern where companies externalize (parts of the) innovation processes to an open innovation marketplace. Companies may be able to draw on a much broader range of ideas and perspectives. They can manage their innovation processes more flexibly and efficiently. At the same time, 'nomadic innovators' bid on innovation tenders and contests in constantly changing teams within this scenario.

In the innovation vision *'deliberative innovation,'* it is expected that citizens will play a more important role both in governing and implementing innovation activities. This innovation model driven by citizens can be best described as a deliberative bottom-up innovation process. The dynamics of cities as hubs and generators for innovation processes was particularly stressed by the related *'city-driven innovation'* vision. In particular, systemic sustainability innovations may best be implemented on a city level in the future.

'Waste-based innovation' explicitly deals with the issue of sustainable production-consumption patterns and the emergence of an innovation model that is fully consistent with a circular flow of resources. McDonough and Braungart (2002) conceptualized this idea as 'cradle-to-cradle principle,' where novelties emerge out of recycled and reused products. The Dutch city of Venlo was selected as a signal as it adopted this idea and wants to radically avoid waste until 2015. However, this was not the only innovation vision that explicitly dealt with sustainability issues. Most innovation visions emphasize the need to address societal challenges and, in particular, environmental issues as a key driver of change, not only for the target or output of innovation but also for the innovation process.

The innovation vision *'negotio-vation'* means that innovation becomes publicly negotiated. Companies, for instance, advertise innovation proposals to citizens and potential customers, who, in turn, vote or approve product proposals and thereby steer the direction of development.

With *'ubiquitous innovation,'* the team expects innovation to become mandatory for more and more people in companies and other types of organizations. The assumption is that the current emphasis on innovation and creativity among designers, programmers, and engineers will spread to all workplaces. Hence, all employees, from the janitor to top managers, are constantly involved in innovation activities. Creativity is part of any daily job routine and is a key in performance measurements.

Despite the overwhelmingly positive association with the innovation visions, we have identified some problematic aspects and effects of these new innovation patterns and even observed some counterbalancing trends, too. The constant pressure to innovate may have negative and unintended consequences: If more and more people suffer from the constant innovation pressure, innovation could become something undesirable and negative in the organization culture. Consequently, people may increasingly feel compelled to use their spare time to meet the never ending innovation demands, which, in turn, could have negative effects on people's health. In addition, professional designers and engineers may feel threatened by the distributed innovation model, fearing job loss due to the large number of new actors who innovate by themselves.

A reverse trend may hence be that innovation fatigue takes over and '*no-innovation*' becomes *en vogue* in certain areas and emerges as a trend in itself. This critical and rather skeptical innovation vision was the only vision among all 20 visions that explicitly expresses a 'deceleration' of the current innovation practice. The 'no-innovation' vision refers to 2 out of the 68 weak signals identified. The first signal is the American Cereal Manufacturer POST, who came up with an anti-innovation campaign concentrating on the message that POST cereals have never changed. This case may indicate a countertrend emerging in some specific product markets where certain products deliberately remain unchanged but are still highly competitive.

The publication of a sustainable design guideline presented at the 'Sustainable Everyday Exhibition' in Milan in 2003 is the second signal identified that served as a base for the no-innovation vision. This guideline formulates in a very simple way what product service designers should be taken into account to align with respective sustainable criteria. The first guideline, for instance, is: 'Think before doing.' Thereby, designers and innovators should weigh up the objectives (because certain design proposals are ethically or environmentally unacceptable in themselves) and their general implications before starting a project. The philosophy is to 'temperate the innovation-oriented mind of the product designers and discourage innovations that are not useful and propose to consider existing artifacts and what can be done with them before inventing new ones.' Thus, a more reflexive way to develop new products is called for, which in specific cases, will slow-down the innovation process.

Within the INFU project, an online survey was launched to assess the 20 innovation visions (for further details see Warnke, Schirrmeister, & Leitner, 2010). In total, 56 experts participated in the survey, the majority of the participants were researchers, innovation consultants, and innovators. The participants were asked to assess the visions with respect to newness, impact, desirability, and likelihood. Concerning the impact of the different innovation visions, 'no-innovation' was considered to have a modest impact in the economic, social, and environmental domain and thereby ranging in the middle field. However, the no-innovation vision was the vision that was considered as the least desirable one, too. In this respect, participants also assessed the 'relocated innovation' vision, 'automated innovation' vision, and the 'negotio-vation' vision as rather undesirable. More than 40% of the participants said that they did not wish that any of those visions became reality. In contrast, 'waste-based innovation' was considered as highly desirable. The 'no-innovation' vision and 'negotio-vation' vision were considered as unlikely, mainly because experts feel that these two visions are slowing down the innovation process. In contrast, the 'ubiquitous innovation' model and the 'innovation marketplace' vision, for instance, were assessed as rather likely. The participants were quite uncertain, however, about the likelihood that a vision becomes mainstream because the spread of assessments was quite large for any of the 20 visions proposed.

In addition, in a workshop involving 12 experts from academia, society, and industry, some positive and negative implications of the 'ubiquitous innovation' vision were discussed in more depth, and the 'no-innovation' vision was considered as a possible counterbalancing trend as well. Although most experts saw some positive impacts of a development toward widespread creativity in all social domains, all in all, it was argued that this extreme vision will hopefully never become reality, as living in a constantly innovating society was considered as nightmare. Among others, the following problematic and negative perspectives were discussed: Despite the fact that innovation will become even more important in the future for all working and living domains, there is a risk that more and more individuals will be part of the innovation imperative without gaining the returns for their efforts to engage in the innovation process. A number of individuals may be competing in crowdsourcing activities under high pressure and low quality working conditions. Thus, at the end, a 'creative poor' class may emerge that cannot benefit adequately from the innovation dynamics and is a subclass of what Florida (2002) described as the creative class. In addition, some experts claim that although the quantity of innovation output has increased, the quality of the output has not necessarily been improved at the same pace. Moreover, we will even see more forms of mental illness, such as burnout, as more and more people will not be able to respond to the rising demands.[3] Thus, in the end, creativity and innovativeness will not pay off for specific societal groups in the long run. The demand of children to be educated in innovation and their involvement in the innovation process may have further negative and unintended consequences that are hard to imagine today.

NO-INNOVATION AS STRATEGY

As described before, the example of a U.S. food company that announced that it is proud of not having changed its products for many years is considered as an indicator of a trend to deliberately reduce innovation activities. Such developments can be linked more generally to a trend in some markets, particularly consumer markets, to sell products that have a high vintage or retro factor. Whereas, for instance, in consumer markets, companies constantly announce new innovations, some firms may actively distance themselves from an innovative image. Instead, they merely focus on their evergreens and rely on old but proven and successful products with a stable market share, for example, Coca Cola.

Within the 'no-innovation' vision, it is assumed that a fast-growing number of consumers deliberately renounces attempts to buy innovative products as they are sick and tired of things constantly changing. Even though companies do not resist innovation completely, they deliberately focus on long innovation cycles by emphasizing the no need for innovation of already 'perfect products.' If such companies innovate, then they hide these efforts to make the products look as if being unchanged (e.g., Rolls-Royce), although innovative technology is behind the dashboard. Such product markets for 'slow down products' particularly exist in consumer markets (living, eating, relaxing, etc.). As a consequence, corporate culture is also changing with less pressure to innovate constantly. Product development which is based on a modularization or platform strategy is another way how companies try to control development and manufacturing costs by using similar components within a series of different products. Offering more product variants is not necessarily related to higher product innovativeness.

How are such innovation strategies assessed by the innovation literature? Traditionally, innovation literature mainly stresses the positive aspects and effects of innovation and largely ignores studying the destructive elements of innovation in more detail, as highlighted already by Schumpeter (1911). Many theoretical and empirical studies have tried to deliver proof that more innovative companies are more successful than less innovative firms. A large part of the innovation literature investigates how firms can speed up the innovation process, for example, by changing organizational structures, introducing innovation management instruments, and so on, and innovation literature studies the reasons for inertia and low innovation performance (e.g., Van Elkan, 1996). Whereas there is literature that investigates under which conditions innovation followers and imitators can successfully compete against innovation leaders, it is mostly assumed that a certain level of innovativeness is required.

However, specific strands of the innovation, economic, and management literature are more critical. Porter (1985), for instance, argues that an innovation differentiation strategy is not necessarily associated with higher economic profits than any other differentiation or cost leadership strategy as long as the strategy is coherent. Thus, for instance, products may not only be differentiated by new product features but also by quality improvements or branding. Innovation literature has examined under which circumstances imitation strategies or second follower strategies are feasible alternatives to a true innovation or pioneer strategy. The empirical findings are mixed with respect to the question whether companies that follow an innovation strategy really achieve a higher performance. The reasons why the results vary in many respects can be miscellaneous: differences in the definition of innovative or noninnovative firms, the sector affiliation of the analyzed firms (e.g., manufacturing versus service firms), no general agreement on how 'performance' is to be defined, and the respective time lags considered between innovative efforts and performance (Brusoni, Cefis, & Orsenigo, 2006).

Lööf and Heshmati (2006) conducted a sensitivity analysis regarding the relationship between innovation[4] and performance using Community Innovation Survey (CIS) data from Sweden. They show that the average innovative firm has higher sales and value added per employee than the average noninnovative firm (for both service and manufacturing firms). However, noninnovative service firms are more profitable than innovative service firms (considering the profit per employee or the sales margin), whereas opposite results were found for manufacturing firms. Regarding growth rates, statistics show differences between manufacturing and service firms. Innovative manufacturing firms have a higher growth rate of value added, sales, and profit per employee. Employment growth is, however, larger for noninnovative firms in the manufacturing sector. Considering firms from the service sector, productivity increases fastest in innovative firms; noninnovative firms in contrast show larger rates of growth in sales and profit.

When differentiating innovative firms as to the degree of novelty of their innovations, Lööf and Heshmati (2006) show a 'closer relationship between innovation output and the level of value added per employee and sales margin' (Lööf & Heshmati, 2006, p. 340). Regarding the relationship between innovation and profit, Cefis and Ciccarelli (2005) show that innovative firms differ from noninnovative firms, insofar as the former are to be found to achieve a higher and steadier level of profitability. Considering firms' growth, a rather different picture is suggested by Brusoni and colleagues (2006), who argue that there is no correlation between innovation and growth indicators. In their meta-analysis, they hence conclude that no systematic effect of innovation behavior on a firm's growth can be found (Brusoni et al., 2006, p. 19).

Moreover, many scholars have claimed that firms tend to reduce competition and hence, if possible, keep competitors with new ideas away from the market. A traditional strategy to cope with the increasing innovation speed and shorter product life cycles is to apply patents that are then not commercialized. Studies show that only about one-third of the applied patents are commercialized (e.g., Webster & Jensen, 2010), and that this share has even decreased in the last few years (Hall & Ziedonis, 2001). Thus, firms increasingly use defensive patenting strategies to gain patent tickets, at least in some sectors, such as the computer hardware industry. In this respect, one can refer to the huge number of papers that study whether and under which conditions the legal protection of ideas encourages or harms innovativeness (e.g., Belleflamme, 2006).

Management literature has argued that companies are very reluctant to be innovative if there is risk to undermine the revenues and profits of their already-existing product lines, a phenomenon that is described as 'product and market cannibalization.' While many scholars have studied how to overcome cannibalization (Buchalter & Sakhnini, 2006), one may also argue that in times of an acceleration of innovation, companies will probably more than ever consider cannibalization as an option. Thus, in the future, companies may even use more and more defensive cannibalization strategies to protect their existing product lines. So-called pseudo-innovation, which is just superficial adaptation, for example, to the product's design, is a further strategy companies can follow to adapt to the constant innovation race.

In this context, it might also be of interest to see whether statistical data about innovation provide evidence of a deceleration of innovation. To address this question, one can refer again to the data from the CIS, the largest survey on innovation behavior of firms conducted regularly across Europe since 1992. The CIS is conducted every four years and covers a three-year time span; it is based on a harmonized questionnaire across all European countries. There are some variables, such as the share of product innovators (= proportion of firms that introduced a new or improved product to the market), which have been included in the questionnaire since 1992, hence a comparison over time can be done for these variables.[5] The CIS data delivers no evidence that the number of product innovators has grown, in some European countries, this number has even declined. On average, 43% of all innovating manufacturing firms in E.U.-15 are product innovators and have hence introduced an improved or a new product on the market within the last three years. The number of product innovators (they are those that have done a product or process innovation according to the survey) has grown only slightly since 1992 and was on average 58% of the firms in the manufacturing sectors in E.U.-15 in 2008.

The CIS collects data about the share of companies that have introduced a new product that is 'new to the firm' and products that are 'new to the market,' whereas the latter are interpreted as having higher innovativeness. This statistic calculated for a set of selected countries reveals no clear trend that would either indicate that the product output and innovativeness (and hence speed) has grown or that innovation is slowing down; the share of both indicators does not vary much between the years (see figure 7.1). The analysis of the turnover achieved with innovative products is another indicator which is of interest (not shown here). The statistic reveals a clearer trend indicating that the importance of product innovation grew between 2000 and 2006 in a number of countries. However, there is one exemption, namely Finland, which shows a reverse trend. This finding may serve as an indication that Finish companies are no longer able to keep up with the innovation pace. This is of particular as interest as Finland usually ranks among the top regarding the CIS but also regarding other innovation indicators.

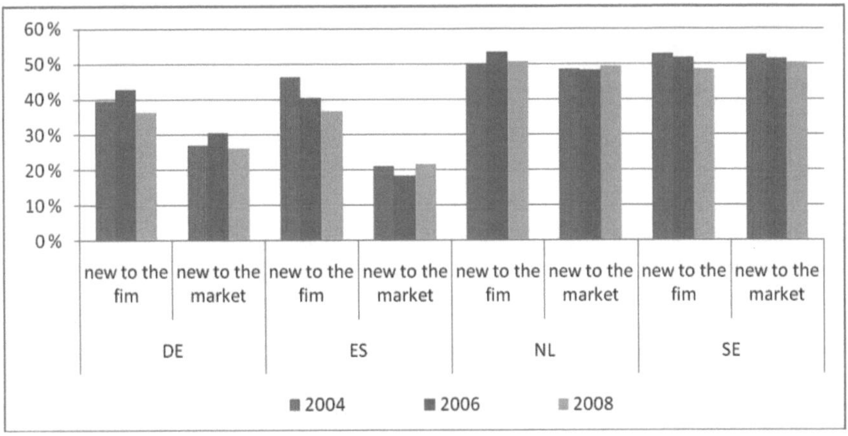

Figure 7.1 Enterprises that introduced new or significantly improved products to the market or to the firm (in %).
Notes
(1) shares refer to enterprises with technological (product or process) innovations;
(2) shares of 2004 are from CIS4 covering all innovations from the years 2002–2004 with 2004 being the reference year, accordingly for CIS2006 and CIS2008;
(3) samples may not be the same for every wave;
(4) the sample covers enterprises from all core NACE activities related to innovation activities.
Sources: Community Innovation Survey (CIS4, CIS2006, CIS2008), Eurostat, own illustration.

More specific evidence of the development of innovation output is hard to find, however. There are hardly any studies available that examine whether the innovativeness of products has grown or whether many product innovations are just incremental adaptations hence resulting in a rise of pseudo-innovations. The increasingly competitive environment in the consumer packaged goods industry, for instance, has led to an increased demand for developing and sustaining new products (Perry & Cochet, 2009). Shorter product life cycles in the consumer market require a constantly growing number of products. At the same time, products flops have grown as well, and only a small part of new products are successful. This, in turn, points to a growing number of innovations (Roth & Sneader, 2006), from which we conclude that, to a larger extent, these might be rather incremental than radical innovations.

To sum up, the traditional sources to analyze innovation dynamics do not deliver evidence of a deceleration of innovation in more general terms. However, they do not indicate that the innovativeness of the outputs has grown either. The lack of measurement options and the underlying definitions regarding the operationalizations of innovation probably explain why we are not able to explore or observe a more differentiated development.

A further interesting indicator to address our question is the development of the number of patent applications. In this respect, statistical data

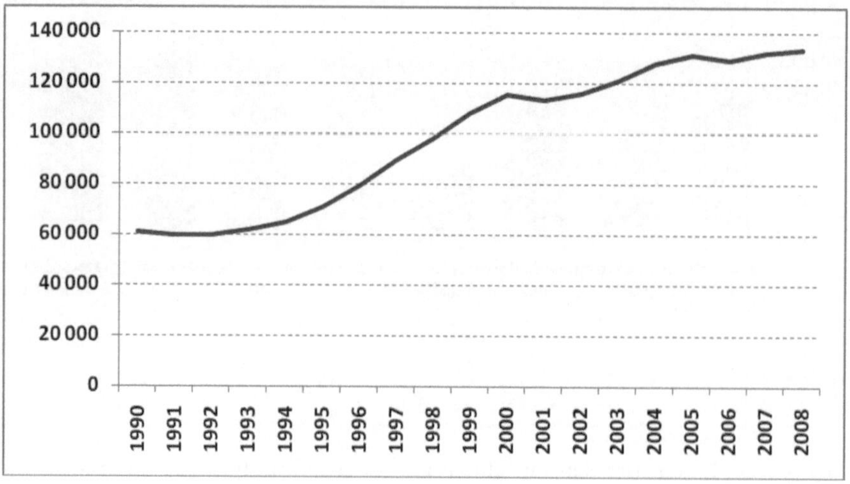

Figure 7.2 Number of patent applications to the EPO by priority year at the national level.
Sources: European Patent Office, author's calculation

reveal that the number of patent applications is growing annually. Although it is well-known that some companies apply for patents that are then never commercialized or that are even strategically applied to reduce innovation competition, there is no evidence that the inventive output of companies and economies has declined. Figure 7.2 reveals the number of patent application at the European Patent Office (EPO). This data shows a relatively strong growth between 1992 and 2000; however, since then, the annual growth rates have declined. An analysis of the World Intellectual Property Organization (WIPO) data shows a similar development, although a slowing down of the number of patent applications cannot be observed in the most recent years (not shown here). However, the trend in increasing patent applications may be driven particularly by the catching up process of different world regions and may slow down in the future as well. Whereas the U.S., Europe, and Japan are still the leading countries regarding innovativeness, many other countries have strongly caught up in the past years. Within the INFU project, some weak signals were identified, revealing that a number of very innovative products and technologies are nowadays developed by companies in countries such as India and China. This development may even culminate in a scenario where the Far East becomes 'innovation leader' whereas Europe and the U.S. will lose their dominant and leading position, as envisaged by the INFU vision 'relocated innovation.' Western scientists and consumers will hence have to accept that they might lose their prominent role in innovation. This shift to the East could lead to less financial capabilities in Western economies for innovation.

We expect that as a result of this changing innovation pattern, some European companies will more often pursue imitation strategies. Thus, although currently product piracy and imitation are said to be mainly done in the new emerging countries, Western countries may start to more often follow imitation strategies in the future. In this context, we have also identified one weak signal pointing to the fact that product piracy and product imitation cases are rising.[6] This signal may also hint at a return to even stricter models of closed innovation processes. The increasing anxiety about product piracy may induce companies to decrease their willingness to integrate external sources into some of the crucial stages within new product development.

Interestingly enough, we can already observe a trend toward a more positive attitude about imitation, which can even be described as a trend to adopt imitation strategies more frequently. In the last few years, a number of business books have been published that advocate the adoption of imitation strategies. Book titles and expressions such as 'copycats' (Shenkar, 2010), 'Imovation,' 'Imitation Economy' (Bennett, 2010), and slogans such as '*Innovation is overrated. It's time to appreciate the power of the copycat*' (Bennett, 2010) are evidence of this trend. Globalization, the modularization of the value chain, the broadening of know-how sharing platforms, and the increased codification of knowledge has made imitation easier than in the past (Shenkar, 2010). Companies will hence more deliberately decide when they innovate and when they want to imitate already-existing products.

However, imitation should not merely be understood as an easy or unreflected activity, but, as authors argue, as an activity that requires a certain level of creativity, very specific capabilities, and, ultimately, a small but certain amount of innovativeness as well. In that sense, imitation becomes a creative process and is not only seen as promising strategy for catching-up countries.

In this context, some researchers have addressed the question of the balance between innovation and imitation. Leijnen and Gabora (2009) have done some research and designed simulation models to study the impact of innovation and imitation. They are modeling societies existing of both innovators and imitators and have found that society is best served when 30% of the population are involved in creating new goods and 70% are focused on copying them.

AIMING AT SUSTAINABILITY AND ITS IMPLICATIONS FOR THE PACE OF INNOVATION

Sustainability research has developed a number of concepts to orient innovation activities toward an economic development with less environmental harm. Among others, the notion of 'eco-innovation' (Bleischwitz et al., 2009) comprises product, process, and system innovations for more sustainability. There is a clear need for eco-innovation for new products and services in fields such as energy, environmental technologies, as well as resource efficiency and globally, 'green markets' are expected to expand heavily. At the same time, the need for systemic eco-innovations underpins the necessity for a transition in established consumption patterns. Hitherto business actors such as the World Economic Forum have pointed out the need for structural transformation of value creation patterns and argues for instance that: 'The world economy must move towards a "new normal," bringing about systemic change in consumption, production and the way in which value is created' (World Economic Forum, 2011, p. 16).

One of the concepts that has become quite prominent relating to the transformation of society and economy toward a more sustainable one is the 'cradle-to-cradle' approach (McDonough & Braungart, 2002). This model implies a radical 'cycling up' approach to innovation and design. According to this model, material flows are designed cyclically as appropriate and continuous biological or technical cycles. All materials at the end of the lifespan of a product are reused and respectively recycled to be productively re-incorporated into new production and product use phases. The cradle-to-cradle design concept is intended to develop highly profitable products whose components are able to circulate in biological and technical cycles with positive effects on the environment.

The vision 'waste-based innovation' in the INFU project adopts the cradle-to-cradle idea and rests on a few weak signals we identified that aim to achieve innovation processes that contribute to sustainable development. The greening and sustainability movement may be seen as a countertrend to the recent innovation race and relentless consumption and may be slowly influencing a culture against too much innovation. The expert group that discussed this innovation vision, for instance, argues that such as an innovation pattern may also be associated with less innovation and consumption. Therefore, a group of citizens or customers not running the innovation race were characterized as 'sleepers.' They are reducing their consumption and spend some of their time in hibernation. 'Many members of society often temporarily opt in for hibernation to reduce cost of living as well as global impact' (Warnke et al., 2010, p. 93).

Another phenomenon, the slow food movement, which has been described as 'voluntary simplicity,' is also related to this trend (e.g., Pietrykowski, 2004; Etzioni, 1998). The slow food movement, which has its origin in Italy, stresses the conscious enjoyment of food and drink and expands the 'concept of consumption beyond that of physical reproduction' (Pietrykowski, 2004, p. 314). The goal of the slow food movement is to create regional networks of agricultural production and consumption that allow small producers to maintain their traditional production methods with other types and less levels of innovation. Such a strategy is clearly related and requires specific social and cultural values. In this respect, Petrini (2001) promotes new forms of 'slow' production and supply that require that consumers value higher quality products with a higher price, however. This trend is also related to the 'slow cities' approach, which is similar to the slow food movement and philosophy. More recently, in urban planning, authors are advocating the creation of places to slow down (e.g., Knox, 2005).

By the same token, 'exnovation' is another concept increasingly discussed in the innovation and sustainability literature. Exnovation is proposed to steer the transformation to a sustainable system that not only requires innovation but also exnovation.[7] Exnovation aims to exclude or destroy everything that is not sustainable. Paech (2005), for instance, argues 'Exnovation is at the end of the innovation life-cycle, where it 'discards' or even purges existing practices to allow the organization to adopt different and fresh thinking to any new innovation activities.' As an example for exnovation, Paech mentions the idea of car sharing. This rather new mobility concept (service innovation) requires that the users refrain from buying a new car after scrapping the old one. In that sense, exnovation is the process of eliminating the unsustainable, and, hence, accomplishes positive sustainability effects that call for an interplay between innovation and exnovation (Fichter, 2010). In the truest sense of the word, exnovation expresses the destructive element of Schumpeter's innovation theory. A more careful development of innovation should therefore foster sustainable development, overcome the dominant growth paradigm, and lead to a slowdown of the pace of innovation.

The discussion on sustainability can also be linked to the critical discourse about economic growth and the role of innovation and technological progress for economic growth (e.g., Paech & von Ossietzky, 2007). Some authors, for instance, claim that a true sustainable development needs to be unlocked from the economic growth paradigm (e.g., Daly, 1996) that, in turn, would require another innovation purpose.

Finally, there is a long history of innovation economics literature stressing that innovation is cyclically driven by technological progress that curb economic development and growth (e.g., Kondratieff and Schumpeter). Phillips (2011), for example, states that at the moment we are in a slowdown phase of such a cyclical up-and-down of technoeconomic progress. Lower incentives to innovate, government regulations that often favor technological incumbents, and some industrials sticking to obsolete business models are identified as reasons for the slowing down of the innovation pace (e.g., Phillips, 2011). However, from a Kontratieff perspective, the next technological revolution is not yet in prospect: the information and communication technology (ICT) cycle is mainly exploited, and nanotechnology as a new cycle does not drive innovation as intensively as initially expected.

OPEN AND DISTRIBUTED FORMS OF INNOVATION AND THEIR IMPACT ON THE INNOVATION SPEED

Most of the innovation patterns identified and discussed within the INFU project are different forms and variants of open and distributed models of organizing innovation involving many different actors, such as suppliers, universities, public agencies, customers, and citizens. The 'negotio-vation vision' and other patterns of innovation visions encompass a high level of participation and involvement by many different actors such as users, citizens, and inventors.

However, although many innovation researchers claim that these new forms of open and user-driven innovation reduce the innovation risk and speed up the innovation processes, some authors also allege that more participatory forms of innovation will slow down the innovation process. An expert group established by the European Commission summarizes highly collaborative forms of innovation as follows: 'The regime of collective experimentation faces challenges because such embedded innovation is laborious, typically loosely-coordinated and slow; as it should be, because users and other stakeholders have their own contexts and logics to consider' (European Commission, 2007, p. 27). Inspired by the 'slow food' movement, one can now proclaim a 'slow innovation' program. Such a slowing down is specifically related to the innovation vision 'negotio-vation' that we have previously presented, assuming that customers and citizens vote by different means relating to the direction of product development. In addition, we assume that in the future, some kind of participation fatigue may emerge that contributes to a deceleration of the innovation process in some areas.

The INFU project also described the visions 'open source society' and 'automated innovation,' which both rest on an extensive use of ICT and expect that we will see more use of ICT for the search for solutions and the identification of customer demands on the web, for web-based product testing, simulations, and so on. Although the intense use of ICT enables new innovation patterns and often speeds up the process, the long-term impact on the innovativeness is hardly ever addressed. Francisco is among those authors who have a very critical perspective on the widespread diffusion and use of ICT in innovation processes. His criticism relates to the automated innovation vision previously mentioned that even proposes and may lead to a development where innovations are only just consumed virtually on the web. Such developments, in turn, may weaken the innovative outcomes of industry in the long run. Francisco argues that large data sources and ICT will crowd out '. . . the opportunity for people to engage knowledge as experience, such as dialoging directly with other people, confronting and creatively solving problems in a socio-cultural context' (Francisco, 2010, p. 4).

Following authors from cultural studies and innovation sociology such as Florida (2002) and Latour (2005), Francisco continues that innovation is a social phenomenon that is created only in the minds and activities of people in a social context. It so far ' . . . relies on language, knowledge, values, internalized experience, an understanding of context and vision of the future—all distinctly human, and distinctly cultivated, characteristics of a social system' (Francisco, 2010, p. 12). Therefore, the increasing use of ICT to bring out innovation and to foster the innovation pace harms the innovative output on the long run. He claims that skills and the changing sociospatial and cultural context to produce innovations is eroding slowly. Francisco (2010, p. 4) describes a negative feedback loop that he labels as 'Innovation Paradox' and defines, 'The more we innovate in our current fashion, the more we disconnect ourselves from the human experiences, behaviors, and values that constitute the roots of innovation as we know it.' Thus, the process of innovation and the way we innovate has a major impact on the innovation output and vice versa.

SUMMARY AND CONCLUSIONS

In economic and business literature there has been a long debate whether companies that innovate are superior to other firms or not. In the last few years, the dominant paradigm expressed by managers, policy makers, and researchers can be summarized as one where innovation is seen as a 'must' in global competition. Researchers have contributed and enforced this development by delivering empirical evidence that innovating firms achieve higher economic performance, such as profit and growth, findings that are the hallmark of policy makers to justify current innovation policies and public investments in R&D and innovation support. The common paradigm in the traditional innovation literature is that more innovation is better than less innovation and that a firm can hardly innovate too much in today's competitive landscape.

However, although the importance of innovation will remain crucial in the future, the question on whether a company, nation, or world region can innovate too much or invest too much in R&D becomes more evident. The INFU project points out that we may see many different patterns of innovation and innovation practices coexisting in the future with the main purpose to raise output and pace of the innovation process. We may in turn, so it is argued in this chapter, see as well some slowdown of innovation in some areas, which, ironically, is an unintended effect of pushing ever forward new innovation models.

A slowing down of innovation may have some positive aspects, too: change in economy structure and focus on innovation in processes to improve quality and efficiency in products. In addition, people enjoy better quality in less 'accelerated' private surroundings as they are steadily forced to be very productive and efficient in their jobs at the same time. The alternative innovation cultures may be accompanied by longer consumption cycles and therefore less waste. The focus on efficient production in these alternative innovation cultures ensures high quality, which, in turn, also ensures high resource efficiency.

While we were not able to observe a broad trend for a slowing down of innovation based on our weak signal identification in the INFU project, there are nevertheless some indications that in specific areas, companies and customers are trying to opt out of the innovation race. To summarize, we propose the following hypotheses for a deceleration of innovation.

First, we will see a slowdown of innovation due to a new trend in deliberately adopting a no-innovation strategy. This may become more relevant for specific industries and markets, particularly in consumer markets (relaxing, eating, living, etc.). This does not mean that companies do not innovate at all, but they innovate generally less by focusing rather on process innovations or by developing products with longer life cycles.

Second, we expect that imitating and copying strategies will become more prominent and professionalized and will be increasingly adopted by Western firms. Copying and imitation strategies may somehow lose its negative image. This development is also related to the visions previously presented as 'relocated innovation' and that have been described as 'reverse innovation' (see Immelt, Govindarajan, & Trimble, 2009) in the literature. If the innovation pace on a global level will increase, and countries such as India and China will catch-up and built up high innovation capacities, innovation competition will become even more harsh, and particularly firms in Western countries may more often follow imitation strategies.

Third, we will see a slowdown due to new participatory and distributed innovation pattern in specific fields. These innovation patterns are explained by larger coordination efforts between different stakeholders to the innovation process, in particular, the end users.

Fourth, as customers and citizens will became more involved and contribute directly in shaping the direction and speed of innovation, we will see a slowdown of innovation in some areas due to higher awareness of customers and citizens concerning the effects of too much innovation on the quality of life and the environment. Green lifestyles and consumption patterns of some consumer groups reward innovation patterns to quality and authenticity embedded in context and tradition.

The conclusion for business managers is to define innovation strategies very carefully by considering the whole range of alternatives, from no-innovation to imitation to fast innovation strategies, and to select the appropriate strategy for different businesses and markets. Thus, companies may adopt an imitation strategy in one business unit while they chose an innovation leadership strategy in another business unit. This obviously requires new types of capabilities and practices for managing tensions within companies. This challenge is similar to the one debated in connection with the ambidextrous organizations (O'Reilly & Tushman, 2008) or mastering the tension between exploration and exploitations in organizational learning (e.g., March, 1991). In addition, companies should more often ask the "why" and "what for" questions related to innovation and thus to more carefully shape the direction of innovation (Butler, 2011).

As innovation processes are embedded in an innovation system consisting of many different actors, such alternative strategies to 'escape' from the intense innovation competition may be difficult to pursue. Due to the current pace of innovation systems, they lose their ability to brake, to stop, or slow down. However, public actors financially supporting innovation processes may decide to support only innovation with a particular value towards certain societal aims, which is one important conclusion for innovation policy makers. In addition, policy makers should more carefully try to shape the direction, purpose and quality of innovation. Hence, economy and society should not innovate just for the sake of innovation but to address societal challenges. However, as previously mentioned and also in other chapters in this volume, policy actors are an element of the innovation system that cannot act independently of other actors and that have difficulties changing established paradigms, too.

Finally, the observations and findings of the INFU project expand our research agenda. Much more research is needed to understand in which context companies should slow down their innovation efforts and what unintended effects on the whole society and economy the still-accelerating innovation pace may have. A slowing down of the innovation pace, for instance, may have a positive aspect on the quality of life but also be effective from a macroperspective as companies will spend resources more efficient.

ACKNOWLEDGEMENTS

This chapter exploits findings from the Innovation Futures project, funded by the 7th E.U. Framework Programme of the European Commission. Fruitful discussions during the project with all partners greatly contributed to the outcome of this research project. Furthermore, my colleagues Claus Seibt, Franziska Kampik, and Marcin Borowiecki have contributed to finalizing this chapter. I am also grateful for helpful comments from the editors of this volume.

NOTES

1. For more information see www.innovation-futures.org. The Innovation Futures project (INFU) has been supported by the 7th E.U. Framework Programme of the European Commission.
2. See also a short video about the 19 innovation visions on the INFU web page, see: http://www.innovation-futures.org/?q=taxonomy/term/51 (retrieval date December 2011).
3. In this respect, one may also refer to the increasing literature in psychology, medicine, and brain research investigating conditions for being creative,

innovative, and healthy. Some researchers, for instance, argue that the information overload today harms creativity, real problem solving, and innovativeness (Interview with Maggie Jackson in *Wired,* February 6, 2009, cited in Francisco, 2010). Moreover, it is well known that great ideas often emerge during leisure time when people are in relaxing environments where time and latitude are given.

4. Lööf and Heshmati (2006) consider a firm to be innovative if its innovation investments are positive and if it has positive innovative sales. Non-innovative firms are firms with (1) neither positive innovation input nor positive innovation output, (2) firms with positive innovation input but no positive innovation output, or (3) firms with no innovation input and positive innovation output.
5. The reader should, however, note that the sample size, response rates, as well as the conducting countries differ from survey to survey. Moreover, the same firms are not necessarily sampled wave after wave (Mairesse & Mohnen, 2010).
6. The weak signal is based on a study of the German Chamber of Commerce concerning product piracy, see http://www.markenpiraterie-apm.de/files/fischerappelt.pdf (retrieval date December 2011).
7. However, exnovation is interpreted and used differently by various authors. Some authors in the field of marketing use the term exnovation to describe innovation processes where the process is initiated from outside (external to) the organization, for example, from customers (e.g., Geoffrey, 2007).

REFERENCES

Belleflamme, P. (2006). Patents and incentives to innovate: Some theoretical and empirical economic evidence. *Ethical Perspectives: Journal of European Ethics Network*, 13(2), 267–288.

Bennett, D. (2010, April 18). The imitation economy; Innovation is overrated. It's time to appreciate the power of the copycat. *TheBoston Globe.*

Bleischwitz, R., Bahn-Walkowiak, B., Irrek, W., Schepelmann, P., Schmidt-Bleek, F., Giljum, S., et al. (2009). Study on eco-innovation—putting the EU on the path to a resource and energy efficient economy. In *Eco-innovation—Putting the EU on the path to a resource and energy efficient economy: Study and briefing notes* (pp. 1–82). Brussels: European Parliament.

Brusoni, S., Cefis, E., & Orsenigo, L. (2006). *Innovate or die? A critical review of the literature on innovation and performance* (CESPRI Working Paper No. 179). Milan: Center for Knowledge, Internationalization and Technology Studies, University of Bocconi.

Buchalter, A., & Sakhnini, H. (2006, February). Fighting cannibalization: Optimization techniques used to plan operations can also be applied to sales and marketing. *McKinsey Quarterly.*

Butler, J. (2011, June 12–15). Innovation frontiers—systems, services, sharing and sustainability. Paper presented at the XXII ISPIM Conference, Hamburg, Germany.

Cefis, E., & Ciccarelli, M. (2005). Profit differentials and innovation.*Economics of Innovation and New Technology*, 14(1/2), 43–61.

Chesbrough, H. W. (2003). *Open innovation: The new imperative for creating and profiting from technology.* Cambridge, MA: Harvard University Press.

Daly, H. E. (1996). *Beyond growth: The economics of sustainable development.* Boston: Beacon Press.

Etzioni, A. (1998). Voluntary simplicity: Characterization, select psychological implications, and societal consequences. *Journal of Economic Psychology*, 19, 619–643.

European Commission. (2008). *The European foresight monitoring network, collection of EFMN Briefs—Part 1.*Directorate-General for Research. Brussels: European Commission.

Fichter, K. (2010). Nachhaltigkeit: Motor für schöpferische Zerstörung?*Soziale Innovation*, 1, 181–198.

Florida, R. (2002). *The rise of the creative class: And how it's transforming work, leisure, community and everyday life.* New York: Basic Books.

Francisco, S. (2010). The innovation paradox: How innovation products threaten the innovation process. *Reconstruction: Studies in Contemporary Culture*, 10(2).

Geffroy, E. K. (2007). *Schneller als der Kunde. Exnovation statt Innovation.* Berlin: Econ Verlag.

Gershenfeld, N. (2005). *FAB—The coming revolution on your desktop—From personal computers to personal fabrication.* New York: Basic Books.

Hall, B., & Ziedonis, R. (2001). The patent paradox revisited: An empirical study of patenting in the semiconductor industry, 1979–1995. *RAND Journal of Economics,* 32(1), 101–128.

Howe, J. (2006). The rise of crowdsourcing.*Wired Magazine*, 14(6). Retrieved November 2009, from http://www.wired.com/wired/archive/14.06/crowds.html

Howe, J. (2008). *Crowdsourcing: Why the power of the crowd is driving the future of business.* New York: Random House.

Hunter, L. (2008). Rethinking production. In G. Gardner & T. Prugh (Eds.), *The Worldwatch Institute: State of the world 2008* (pp. 38–40). Washington, DC: Worldwatch Institute.

Immelt, J., Govindarajan, V., & Trimble, C. (2009). How GE is disrupting itself. *Harvard Business Review,* 87(10), 56–66.

Jégou, F., Leitner, K.-H., Mahn, J., Rhomberg, W., von Saldern, S., et al. (2009). *Structured collection of current signals for arising changes in innovation patterns.* Deliverable 1.1 of the project 'INFU—A Foresight Exercise on Emerging Patterns of Innovation,' Project funded within the 7th Framework Programme of the European Commission.

Jégou, F., Leitner, K.-H., Mahn, J., Mueller, M., Pitsci, G., Rhomberg, W., et al. (2010). *Final set of 20 amplified and contrasted visions.* Deliverable 2.3 of the project 'INFU—A Foresight Exercise on Emerging Patterns of Innovation,' Project funded within the 7th Framework Programme of the European Commission.

Kim, W. C., & Mauborgne, R. (1999, September–October).Strategy, value innovation, and the knowledge economy. *Sloan Management Review,* 78, 129–141.

Knox, P. (2005). Creating ordinary places: Slow cities in a fast world.*Journal of Urban Design*, 10(1), 1–11.

Latour, B. (2005). *Reassembling the social: An introduction to actor-network-theory.* New York: Oxford University Press.

Leijnen, S., & Gabora, L. (2009). How creative should creators be to optimize the evolution of ideas? A computational model. *Electronic Proceedings in Theoretical Computer Science*, 9, 108–119.

Lööf, H., & Heshmati, A. (2006). On the relationship between innovation and performance: A sensitivity analysis. *Economics of Innovation and New Technology*, 15(4/5), 317–344.

Mairesse, J., & Mohnen, P. (2010). Using innovation surveys for econometric analysis.In B. H. Hall, & N. Rosenberg (Eds.), *Economics of innovation* (vol. 2, pp. 1129–1155). Amsterdam: North Holland.

Manzini, E., & Jegou, F. (2003). *Sustainable everyday—Scenarios of urban life.* Milan: EdizioniAmbiente.

March, J. G. (1991). Exploration and exploitation in organizational learning. *OrganizationScience*, 2(1), 71–87.

McDonough, W., & Braungart, M. (2002).*Cradle to cradle: Remaking the way we make things.* New York: North Point Press.

Miles, I. (2008). Foresight methodology. In L. Georghiou, J. C. Harper, M. Keenan, I. Miles, & R. Popper (Eds.), *The handbook of technology foresight: concepts and practice.*Northampton, MA: Edward Elgar.

Mulgan, G., Ali, R., Halkett, R., & Sanders, B. (2007). *In and out of sync: The challenge of growing social innovations* (NESTA Research Report). London: National Endowment for Science, Technology and the Arts.

Mumford, M. D. (2002). Social innovation: Ten cases from Benjamin Franklin. *Creativity Research Journal*, 14(2), 253–266.

O'Reilly, III, C. A., & Tushman, M. L. (2008). Ambidexterity as a dynamic capability: Resolving the innovator's dilemma. *Science Direct*, 30(28), 185–206.

Paech, N. (2005). *Nachhaltiges Wirtschaften jenseits von Innovationsorientierung und Wachstum. Eine unternehmensbezogene Transformationstheorie.* Oldenburg, Germany: Metropolis.

Paech, N., & von Ossietzky, C. (2007). *Economic growth, environmental innovation and the dematerialization myth.*oikos Conference 2007, St. Gallen, Switzerland.

Perry, C., &Cochet, M. (2009). Consumer packaged goods product development processes in the 21st century: Product lifecycle management emerges as a key innovation driver. In H. R. Moskowitz, I. S. Saguy, & T. Straus (Eds.), *An integrated approach to new food product development* (pp. 113–131). London: Taylor and Francis.

Petrini, C. (2001). *Slow food: The case for taste.* New York: Columbia University Press.

Phillips, F. (2011). The state of technological and societal change: Impressions. *Technological Forecasting and Social Change*, 78, 1072–1078.

Pietrykowski, B. (2004). You are what you eat: The social economy of the slow food movement. *Review of Social Economy*, 62(3), 307–321.

Pollock, R. (2006). *Innovation and imitation with and without intellectual property rights* (MPRA Paper No. 5025, Cambridge University). Retrieved December 2011 from http://mpra.ub.uni-muenchen.de/5025

Porter, M. E. (1985). *Competitive advantage: Creating and sustaining superior performance.* New York: Free Press.

Roth, E., & Sneader, K. D. (2006, November).Reinventing innovation at consumer goods companies.*The McKinsey Quarterly.*

Schumpeter, A. J. (1911). *Theorie der wirtschaftlichen Entwicklung.* Berlin: Dunker &Humblot.

Shenkar, O. (2010). Copycats: How smart companies use imitation to gain a strategic edge. *Strategic Direction*, 26(10), 3–5.

Stoneman, P. (2007). *An introduction to the definition and measurement of soft innovation* (Working Paper). London: National Endowment for Science, Technology and the Arts.

Stoneman, P. (2009). *Soft innovation—Towards a more complete picture of innovative change.* London: National Endowment for Science, Technology and the Arts.

Tuomi, I. (2002). *Networks of innovation: Change and meaning in the age of the Internet.* Oxford: Oxford University Press.

Van Elkan, R. (1996). Catching up and slowing down: Learning and growth patterns on an open economy. *Journal of International Economics*, 41(1–2), 95–112.

vonHippel, E. (2005). *Democratizing innovation*. Cambridge, MA: MIT Press.
Warnke, P., Schirrmeister, E., & Leitner, K.-H. (2010). *Innovation futures scripts: Nodes of change in innovation patterns emerging from the explorative dialogue on the 19 INFU Visions*. Deliverable 3.1 of the project 'INFU—A Foresight Exercise on Emerging Patterns of Innovation,' Project funded within the 7th Framework Programme of the European Commission.
Webster, E., & Jensen, P. H. (2010). *Do patents matter for commercialization?* (Working Paper No.8/09, Melbourne Institute of Applied Economic and Social Research). Melbourne: University of Melbourne.

Part III

Exploring Unintended Consequences of Innovation

8 Do Major Innovation Models Consider Unintended Consequences?
A Review and Revised Framework

Martin Lindell

The fundamental impulse that sets and keeps the capitalist engine in motion evolves from new customer goods, new methods of production or transportation, and new markets and/or new forms of industrial organizations that capitalist enterprises create (Schumpeter, 1959). Innovations are considered a positive activity and a driver of development in society. However, at the same time, Schumpeter admits that innovations may also have negative effects, as they may destroy an earlier process, product, or system.

Innovations affect numerous stakeholders, including society, consumers, and employees. Traditionally, the focus in developing innovations has been on the needs and interests of the customer, the top management, and, at times, the shareholders, whereas little or no attention has been given to the interests of employees and society.

The aim of this chapter is to analyze the application of innovation models from the perspective of employees and society with a particular focus on the negative consequences that innovation processes may create for these stakeholders. Furthermore, this chapter examines the means to persuade developers to pay greater attention to the negative consequences of innovations on employees and society.

Innovation is discussed in several contexts, including product innovation, process innovation, service innovation, and organizational innovation. With respect to product innovation, we refer to those changes (often radical) in products that a firm offers. Process innovations are changes to the way the products are produced and delivered. Service innovation includes a generation of new service and organizational concepts intended to promote company creativity in a turbulent environment. In this chapter, we focus on product and process innovation and the progression of innovation over time.

A review of a number of influential innovation models in the innovation management literature is conducted, and we investigate, especially, the content and possible negative effects of the internal developmental work. The review concludes with a classification of the models based on their main content and purpose. Because the models do not include an explicit discussion of the negative effects of innovation, a section of examples is included.

In this section, negative effects of innovation, both for employees and for society, are described. Throughout this chapter, we argue that the innovation models must address the negative effects for all relevant groups inside and outside the companies, thus focusing on the stakeholder. The focus must include more than just a perspective from upper-level management that stresses technical and commercial success. The models must include the activities after the launching of the innovations, that is, the implementation, the use, and even the retiring of innovations.

A REVIEW OF INNOVATION MODELS

This review is focused on four different types of models. These include product and process development models, market acceptance models, competition and life cycle models, and management and growth models. The product development models are the oldest ones. For the others, the 1980s was a very active decade.

The Planning Approach and Related Models

To gain an understanding of agents and factors that have an impact on innovation, we must have a solid understanding of the innovation development processes. Existing models in the early innovation literature have been classified in different ways. Two common ways to divide innovation processes are the *planning models and the emergent innovation models*. The planning models typically break down the innovation process into the following six steps (cf., e.g., Cooper & Kleinschmidt, 1986; Johne & Snelson, 1988; Robertson, 1967; Urban & Hauser, 1980; Zirger & Maidique, 1990):

1) The definition of product specifications
2) The development of alternative concepts
3) The identification or creation of testing materials, components, and constructions
4) The development of prototypes and field tests
5) The adaptation to production and marketing demands
6) The launch of the innovation

Many researchers have criticized the planning approach and stress the emergent character of innovation development processes as they analogize that invention and innovation are akin to exploring an unfamiliar coastline in the fog (Schon, 1967). That is, the development process of an innovation has an emergent character where there is much experimentation and learning throughout the process. Accordingly, actions are taken in small steps, and evaluation and decisions are made continuously throughout the process.

A third classification of innovation models is based on the *main source or actors for ideas* for the development process. Some innovation models focus on analyses of customers and the use of new products (need-pull models, cf. Burgelman & Sayles, 1985). A second way to explore innovation is from a so-called technology-push approach, where research and development (R&D) managers and scientists have the dominating role. A third approach could be called the champion approach (champion driven), an approach that explores innovation development as so-called skunk work. In such an approach, individual managers, often at the bottom of the organization, push to adopt a new business innovation in a larger organization without approval from top management. Thus, managers that could be called project or organizational champions have the main role (Burgelman, 1983).

The previous models cover the development from the inception of the idea to the launching of the innovation. All of these models thus have a primarily internal focus, as they are all concentrated on the efforts to accomplish the development of a successful product based on customer demand (Bessant & Tidd, 2007; Cooper, 2001; Langrish, Gibbons, Evans, & Jevons, 1972; Rothwell, 1972; Zirger & Maidique, 1990). No analyses of the consequences of innovations for the companies or for the greater society are included in these models.

The next step is to consider some postlaunching models. The market diffusion model, for example, is focused on the spread of the innovation on a market. The growth and product and life cycle models describe the demand of innovations over a life cycle. And, finally, different growth models outline the supporting organizational context for an innovative growing firm.

Market Diffusion Models

Since the early 1960s, the leading researcher on innovation diffusion has been Everett Rogers. He is well known for his creation of the adoption curve (Rogers, 1995), whereby the customers are divided into five categories: innovators, early adopters, early majority, late majority, and laggards. The transition from one group of adaptors to another is not always easy to identify or easy to accomplish. Moore (2001) speaks of a chasm between the groups. Rogers, aware that innovation might have effects both at the individual level and at the level of the social system, discusses the desirable and undesirable consequences that may arise depending on whether the effects within the social system are functional or dysfunctional. However, the negative effects are not more considered in his adoption curve. See chapter 5 in this volume for a discussion Rogers's model at length.

Competition, Product Life Cycle, Management, and Growth Models

The many types of models describing the life cycle trajectory of innovations or technologies have been separated into two groups. The first group focuses on marketing, technological, and competition issues. This group consists of four types of models: the technological life cycle model, the technology s-curve model, the product-process model and the dominant design model. The *technology life cycle model* maps the sales volume of a technology over time. Levitt (1965) divides the growth into the following four traditional phases: market development, market growth, market maturity, and market decline. He further emphasizes the need to begin planning for the s-curve in the development phase. The *technology s-curve model* maps the technical performance trajectory of a technology in relation to (R&D) effort (Foster, 1986). The general idea is that the return on R&D investment varies. For example, when initially developing a new technology, one must invest proportionally more, and in the later maturing phase, one might receive proportionally less return on investments.

The product-process cycle maps product and process innovations within a sector over time (Abernathy & Utterback, 1978). In the beginning, the focus is on product innovations, but in a more mature industry phase, the focus transitions to production innovations. To be successful, an integration of marketing, production, and R&D development is essential (Etienne, 1981). The *dominant design model* maps the emergence of a dominant design of an innovation or technology over time. There are currently several competing technological designs on the market (Gardiner & Rothwell, 1985) as technology tends to evolve over relatively long periods with incremental changes that are punctuated by a relatively rare new winning technology (Tushman & Anderson, 1986). All of these models describe the demand of an innovation over time and the competitive action taken by management to increase and extend that demand with a focus on technology development, competitive actions, consumer behavior, and decision making rather than on the consequences of the innovations themselves.

The second group of growth models consists of those focusing on the internal organization, especially on *the administrative systems and organizational structure* necessary to support the growth of the business. Moore and Tushman (1982) describe "managing innovation over the product life cycle" as they break the growth process into the introduction phase, the growth phase and the mature phase, whereas Adizes (1988) has identified 10 phases. However, again, nothing is mentioned about possible side effects that innovations might have during their different phases. Instead, the authors note that the vast majority of failing innovations fail due to marketing and organizational pathologies rather than the lack of usable technologies. Whereas many small firms fail because they are unable to make the transition to more mature larger organizations, many large firms also fail as they become more and more bureaucratic, thus stifling their ability to innovate.

Miller and Friesen (1983) have noticed that during the growth of a corporate life cycle, successful development shows gradual and continual increases in the sophistication of their information processes and decision-

Table 8.1 Summary of Models of Innovations, Product Development and Growth and Researchers Associated with Each Model

Models	Key content	Researchers
1. Innovation and produc development models	Focus is on the development phase, that is, to get a product on the market. Success and failure of new products are analyzed. The organization of R&D activities is outlined and described; the competitors and completion are investigated. In the most recent studies, possible negative effects, postlaunch, and product retirement are mentioned for the first time.	Rothwell (1972) Langrish et al. (1972) Zirger & Maidique (1990) Cooper (2001) Bessant & Tidd (2007) Ahmed & Shepherd (2010)
2. The product-process model	Change of focus in the development phase from product to process, with a R&D focus.	Abernathy & Utterback (1978) Etienne (1981) Rogers (1995) Moore (2001)
3. Market diffusion models	Focus on the acceptance rate and diffusion process. Possible negative effects explain slow diffusion, not viewed as negative phenomena as such.	Levitt (1965) Foster (1986)
4. Technology life cycle and technology s-curve models	Prediction and description of the demand of new technologies and products. Stress on competiveness.	Gardiner & Rothwell (1985)
5. The dominant design model	The winning design among a number of competing alternatives. A technology focus.	Tushman & Anderson (1986)
6. Growth and product life-cycle models	Stress on management and organizational effects of growth. Different growth phases are carefully described. No attention to the negative effects of innovations.	Greiner (1972) Miller & Friesen (1983) Moore & Tushman (1982) Adizes (1988)

making methods. Organizational characteristics and individual behaviors that are vital at time one might be absolutely wrong and dysfunctional at another time, in which case the whole organizational configuration must be transformed. Greiner (1972) describes how growing organizations, over time, move through periods of stability punctuated by periods of instability.

According to Moore and Tushman (1982), top management of successful companies must balance between several paradoxes, such as stability and change, consistency and inconsistency, centralization and decentralization, and order and disorder. Those firms that can master these paradoxes over time will be among the most successful. Moore and Tushman (1982) further conclude that the two most critical management problems are (1) managing the transition from a successful entrepreneurial firm to an established, functionally oriented company and (2) retaining the ability to learn and innovate during the mature phase.

In addition to focusing on the direct modeling of development processes, researchers have devoted considerable effort to understanding life cycles and competitive actions, although management and organizational issues have also received substantial attention.

Summary of the Innovation Models

Table 8.1 provides a summary of the previously mentioned product and innovation development models, along with specific authors associated with each type of model. Regardless of the classification or the models used, however, the negative consequences for employees and/or society are rarely taken into consideration. This is especially true for those models that focus on the development process. Only in some of the most recent literature on innovation processes are the negative effects for society even briefly addressed (cf. Ahmed & Shepherd, 2010).

Why do the models not address the negative effects? One of the main reasons seems to be that the developers are focused only on the technological and market side of the development work, a focus that is far too narrow. In fact, the developers do not consider the changes needed within the organization or in society when innovations are implemented. Bruce (1993) strongly emphasizes that there is a difference between what the developers of an innovation intend and what actually occurs when the innovation is realized in a particular social setting. That is, there is a significant difference between the ideal and the real. According to Bruce, the developers may intend for the innovation to modify the social system such that certain desirable improvements are achieved. However, that process is, in truth, idealization. In fact, the social system may or may not change at all, and if it does change, it may not do so in accordance with the goals and intent of the developer, or there may be a range of changes that occur both in the innovation phase and within the context of use/implementation whereby each change triggers another change.

An interesting phenomenon is that new concepts and actions that at least partly aim to correct the negative consequences of earlier innovations have recently emerged in the literature and in society. For example, Bessard and Tidd (2007) introduce the concept social entrepreneurship, which is a form of innovation that pursues mainly social goals (Peredo & McLean, 2006;

Zahra, Gedajlovic, Neubaum, & Shulman, 2009). It is these social entrepreneurs who attempt to solve large social problems and promote social change by employing traditional entrepreneurial principles to organize, create and manage a venture (Mair & Marti, 2006). Primarily non-profit-oriented, their goal is to generate social value.

In the same spirit, Ahmed and Shepherd (2010) talk about social innovation, a phenomenon characterized by environmental lobbying and pressures, such as the lobbying for stricter controls on pollution, as in the Kyoto Protocol. They also discuss political innovations, innovations that manifest themselves as institutional reforms, social direction, and governance.

RESEARCH AND EXAMPLES OF NEGATIVE CONSEQUENCES OF INNOVATIONS FOR EMPLOYEES AND SOCIETY

In the previous sections, we indicated that there might be some negative effects of innovations on employees and/or on society. However, we have found no in-depth discussions about the negative effects of innovation in the previously described models. Is that because there are no such effects? A literature search was conducted to find articles that focused on the effects of innovations on employees and society. However, these studies were not connected to any innovation process models but were separate studies on the consequences of innovation on employees and society. In the following are some reported examples that describe the negative effects of innovations.

Innovation and Negative Consequences for Employee Well-Being

Some researchers have concentrated on whether innovations create or destroy jobs. Pianta (2006) emphasizes that the radical innovations in computers and telecommunications have led to rapid improvements in technologies with widespread application possibilities that both create and destroy a significant number of jobs. The general expectation is that innovations, especially in production processes, will reduce the number of available jobs. Zimmermann (1991), for example, in a study of German firms, found that technical advances had a negative effect on employment. In another extensive German study, Peters (2005) noticed that manufacturing firms that solely carried out process innovations also experienced negative employment effects, especially those firms that concentrated on rationalization innovations.

On the other hand, Benavente and Lauterbach (2007) conducted a study in a Chilean manufacturing sector, and they could not identify any negative effects of innovation on employment. Similarly, Lachenmaier and Rottman (2007), in a study of the manufacturing sector in Germany, found no negative effects of innovation on employment. After an extensive review of the literature on innovation and employment, Pianta (2006) determined that although there were a significant number of studies that indicated a negative impact of innovation on employment, there were also a similar number of studies that indicated a positive impact. Innovation appears to have a net job-creating effect in those manufacturing and service industries showing high growth in demand and an orientation toward product innovation, whereas new processes appear to result in job losses (Pianta, 2006).

At the company level, Harrison and colleagues (2008) also found that process innovations lead to job destruction, particularly in the short term, but that this is often compensated for by the employment growth that results from product and service innovation. So, all in all, new work methods or processes and new technology can be sources of labor intensification, and although the changes in employment numbers might not be that radical, the changes in the content or type of jobs are significant and can lead to stress and anxiety.

Bryson and colleagues (2009) investigated the effects of managerial innovations on the well-being of employees. According to their findings, innovation can be met with resistance or hostility by employees who are either fearful of change or believe that the change will in some way harm them. Zwick (2002) found that innovation renders current knowledge obsolete, thus forcing employees to undergo training to work efficiently once the innovations are implemented. In addition, innovative working practices (IWP) are often introduced as part of broader structural and organizational changes, such as the introduction of a new plant or technology, changes in management structure, or the introduction or redesign of products and services. Such innovations may also bring with them the threat of job loss resulting in job insecurity, which, in turn is associated with a negative effect on the worker's well-being (De Witte, 1999). According to Harrison and Legendre (2003), decentralization, the involvement of the employees, and the introduction of modern management can likely counteract the possible negative effects of innovations.

Kalmi and Kauhanen (2005), researching the impact of workplace innovation on employees in Finland, concluded that there are two different views on the impact of workplace innovation on employees. The first strand of the literature suggests that employees will benefit from innovative practices (Appelbaum, Baily, Berg, & Kalleberg, 2000; Freeman & Kleiner, 2000). The other strand is more critical and more negative with respect to the benefits for employees. Ramsay and colleagues (2000) claim that top management adopts workplace innovations to increase profits and productivity, whereas the employees bear the cost through increased work intensity, increased job strain, greater stress, and reduced satisfaction. Kalmi and Kauhanen's own study (2005) was more in alignment with the former perspective, which found that employees tended to benefit from workplace innovations.

Kanter (1985) addresses the tension that often exists when developing existing businesses. This tension might, to some extent, explain the resistance of employees toward innovations and the increased stress that employees often experience. According to Kanter (1985), innovation and new ventures have four particular characteristics that account for their special management requirements: uncertainty, knowledge-intensity, competition with alternative courses of action, and boundary crossing. Whereas entrepreneurial management places significant emphasis on such features as visionary leadership, patient money, planning flexibility, team continuity/stability, and interfunctional cooperation, the usual requirements of administrative management in established corporations contradict these principles. Accordingly, Kanter (1985) posits that all companies need to not only manage ongoing activities but to create new ones. There is, therefore, a need to strike a balance between administrative and entrepreneurial management.

Innovation and Negative Consequences for Society

Innovation may generate not only problems for the employees of the companies that implement them but also for the whole of society. Some innovations have increased the possibilities for accidents for the user of the innovations. Consider, for example, the statistics from the U.S. that clearly demonstrate cell phone use contributes to car accidents. In 2009, in the U.S., 995 people died and 24,000 were injured in car accidents where cell phone use was identified as a contributing factor. Stayer and colleagues (2006) determined that the effect of using a cell phone when driving can be as profound as driving whereas drunk (0.08% blood alcohol concentration). When drivers were using either a handheld or a hands-free cell phone, their reaction times were delayed, and they were involved in more traffic accidents than were drivers who did not use a cell phone when driving. Cell phones have also changed the behaviors of the consumers. For instance, in schools and universities, students use cell phones during lectures and classes, thus diminishing the learning experience in the classroom.

Society is also confronted with a significant waste problem, an issue that has not been solved by the innovators, at least not in the short term. In the pulp and paper industry, this problem is evident in the dioxin that is released by bleaching with chlorine, whereas in the case of dry cell batteries, cadmium, mercury, lead, nickel, cobalt, lithium, and zinc are either contaminating the landfills or being released into the air (Porter & Van der Linde, 1995). These negative effects of innovations are issues that plague society and must be addressed. This is especially true for telecommunication devices such as computers and cell phones that require special waste treatment methodologies. Because the innovators did not consider the end results—what to do with the product once it is no longer useful—society is threatened by an ever-growing pile of waste.

In summary, we can conclude that innovation impacts both employees of the respective companies and society as a whole and that the impact is not always positive. It is not easy, however, to explain why these negative aspects have not been addressed in earlier innovation models.

DISCUSSION

Our review of innovation models clearly indicates that innovation models have only, at best, minimally considered the negative effects of innovation. In this section, we first discuss how the models could be improved. Second, we suggest some compulsory measures for correcting the negative effects of innovation. Third, we suggest improvements for management in innovative companies. Finally, we conclude by presenting our own broad framework for developing innovation models in the future.

Model Improvements

It is clear that the effects of innovation on employees and society must be considered more explicitly in the future. Bessard and Tidd (2007) note that a company can only secure a license to operate if the company can demonstrate some concern for the wider community in which it is located. According to Bessard and Tidd (2007, p. 325), innovation must solve various types of environmental problems such as:

- Cleaner products—products with a lower environmental impact over their life cycle
- More efficient processes—processes that minimize or treat waste, reuse, or recycle
- Alternative technologies—technologies that reduce emissions and provide renewable energy
- New services—services that replace or reduce the consumption of products
- Systems innovations—innovations that measure and monitor environmental impact, new sociotechnical systems

The largest potential for improvement appears to be in the R&D process models (the first type of models mentioned in table 8.1). The traditional product development model ends with the launching of the product, but the negative effects do not appear until the implementation and use of the innovation. Therefore, the process models must be extended so that they incorporate the implementation and use phase of innovations. Ahmed and Shepherd (2010), for example, include postlaunch and product retirement as a final last step in their model of innovation processes. Developers should have growing concerns for the environment and should promote the enactment of laws intended to constrain pollution and protect the ecosystem. For example, product developers should explicitly consider the environmentally safe disposal of a product when it no longer serves a purpose. Recycling and disposal plans are essential. Furthermore, the existing models can, to some extent, be improved by introducing more testing of the models, especially during the implementation and use phase. It is also suggested that human resource and social responsibility components be promoted during the implementation and use phase, as these components are currently lacking in the models. Such a module could eliminate some of the negative consequences both for employees and society.

The structure of the other models in table 8.1 appears to be quite satisfactory. They have been developed to illustrate specific phenomena, such as market introduction, life cycle, and growth of innovation. For these models, it is more a question of developing the content of different phases than to improve their structure. Specifically, the growth and life cycle models (type 6 models in table 8.1) could very well be extended by incorporating an elaborated management function based on the action plan we suggest later in this section.

Stakeholders Enforcing Restrictions on Innovation

If the companies themselves fail to correct for the negative effects of innovations, stakeholders outside the companies may feel an obligation to respond. Legislators can force innovators to assume greater responsibility for the negative effects by enforcing specific laws and regulations. Legislation is particularly common in the field of medical innovation and is becoming more common for environmental protection. Trade unions can (and have) also put pressure on companies and top management to pay more attention to the negative effects of their innovations.

Earlier studies indicate that legislation is the strongest measure (cf. a Nordic study by Ytterhus & Synnestvedt, 1995). Nordic managers seem to have a more positive attitude toward restrictions through legislation than, for instance, U.S. managers with respect to environmental investments. Resolving the negative effects caused by innovations appears to be dependent on the interactions among government, customers, suppliers, owners, managers, and employees of the respective companies. As improvements in reducing pollution require interaction and cooperation among all groups, at least in the long run, environmental improvements can go hand-in-hand with business competitiveness and profits.

Top Management Actions

To achieve lasting change and to most effectively eliminate negative consequences, the companies as well as their top management, developers, and employees must change their behaviors. One basic problem seems to center around management philosophy and short-term profit orientation. Conway and Steward (2009) underline the problems with the measures that have been used by top management. Success, for example, has been measured by *financial criteria*, such as the level of profit or turnover generated by the innovation or the speed of the return on the investment (ROI); by *market criteria*, such as the rate of adoption of the innovation; by *technical criteria*, such as the elegance of engineering design or improvement in performance and functionality; by *strategic criteria*, such as the building or sustaining of competitive advantage through the development of superior product or service offerings or the building of technical competences or capabilities; or by *process criteria*, such as the compression of the time required from idea conception to market launch (Conway & Steward, 2009, p. 24). The measures used to determine success have been based on shareholder view and/or on profit maximization, whereas the negative effects of the innovations have been ignored. Furthermore, profit maximization is only a partial valuation and is, therefore, an inadequate reflection of a company's real ability to create value.

Adopting a stakeholder view would result in a change of thinking and values. From such a perspective, managers of companies serve as agents of the stakeholders and have two responsibilities (Smith, 2003). Their first responsibility is to ensure that the ethical rights of the stakeholder are not violated. Their second responsibility is to balance the legitimate interests of the stakeholder when making decisions. The most important stakeholders are the shareholders, the employees, and the customers, and each group makes a significant contribution in that shareholders commit their capital; employees provide their expertise, effort, and time; and customers expend their resources to buy a company's product or services. Thus, new measures of success that embrace the well-being of the employees and society are needed. Conway and Steward (2009) note that the success of an innovation could be judged by the impacts or the consequences that the innovation brings through its usage to individual users or, although its wide diffusion, to society. The question is twofold: How do we convince companies to seriously consider the negative effects of innovations on employees and society, and how do we transform companies' orientation from the short term to the long term?

Ellsworth (2002) strongly emphasizes that the purpose of the company is critical, suggesting that the remedy begins with a purpose that defines ends that people find meaningful and worthy of their commitment. Company management must define the basic end values for which the company stands. Strategies based on short-term returns to shareholders erode the company's competitiveness and, thus, its ability to provide long-term returns to all stakeholders, that is, shareholders, employees, customers, suppliers, and society. A strong, short-term profit orientation in companies often leads to conflicts of interest and adversarial relationships (Smith, 2003).

From a long-term perspective, the most important stakeholder group is the customers, followed by the employees (Ellsworth, 2002). The primary focus should be on serving and meeting the customers' interests and needs followed by addressing the needs of the employees. Kotter and Heskett (1992) have found that companies with a strong culture and customer orientation tend to generate the best total returns for their shareholders. When determining and adopting the overarching purpose(s) for a corporation, the following qualifications should be considered (Ellsworth, 2002, p. 6):

- provide the greatest focus on achieving a competitive advantage
- lay the best foundation for achieving harmony among purpose, strategy, goals, and shared values
- motivate managers to create the maximum level of total value, not just value for one particular constituent
- heighten the motivation of members of the organization, subordinating narrow self-interest to corporate ends, and enhancing the firm's ability to create knowledge
- be seen by employees around the globe as a worthy and meaningful end to serve and serving this end, thus potentially raising employees' moral aspirations
- promise to enhance the legitimacy of the corporation's actions in society

The leader/top management is responsible for defining a purpose that has a significant positive effect on corporate performance, on employees, on the satisfaction of human needs, and, thus, on society. In this initial phase of defining the company's purpose and the purpose of the innovation, it is also possible to consider the negative consequences of the implementation. Defining the purpose and establishing priorities for various stakeholders demands that top management consider and evaluate the following three points (Ellsworth, 2002, p. 42):

1) Which of the constituents'/stakeholders' interests best serves society?
2) Which constituents/stakeholders have the greatest impact on the firm's wealth-producing capacity?
3) Which specific purpose results in the best competitive performance?

Establishing a well-defined purpose that is the result of in-depth discussion creates a positive climate in the organization, releases built-up energy, and eliminates many of the possible negative effects for both the organization and society.

AN AMENDED FRAMEWORK FOR DEVELOPING INNOVATIONS

Based on the previous ideas, a framework has been developed that might minimize the negative effects of innovation. Under this new framework, the core development process is extended to include the implementation phase of the innovation as well as diffusion, growth, and product retirement phases. Therefore, the developers and the developing companies have a responsibility to follow-up after the implementation and the use of the innovation and take necessary corrective actions. The wishes and demands from the different stakeholders, society, customers, and employees are the basis of the development activities. The central actors are the developers and top management, with the most critical role being assumed by top management, as it is the responsibility of top management to define the purpose, goals, and strategies for the company in such a way that possible negative effects of innovations are considered in the early phases and, thus, minimized. In this way, the interests of all stakeholders are considered. Top management, however, must take the supervisory role more seriously and be more active in screening ideas and projects in different phases of development. Furthermore, top managers should demand more comprehensive evaluations for innovation projects, including scenarios for possible unintended and undesirable consequences for employees and society.

The innovation developers must have a broader perspective of the innovation processes. That is, they must be aware of the possible negative effects of the innovation. It is not enough to launch an innovation and thus meet the demands of the market, as the innovation cannot be deemed successful until all negative effects have been identified and corrective measures taken to resolve the problems. The outline of a new framework that takes these factors into account is presented in figure 8.1.

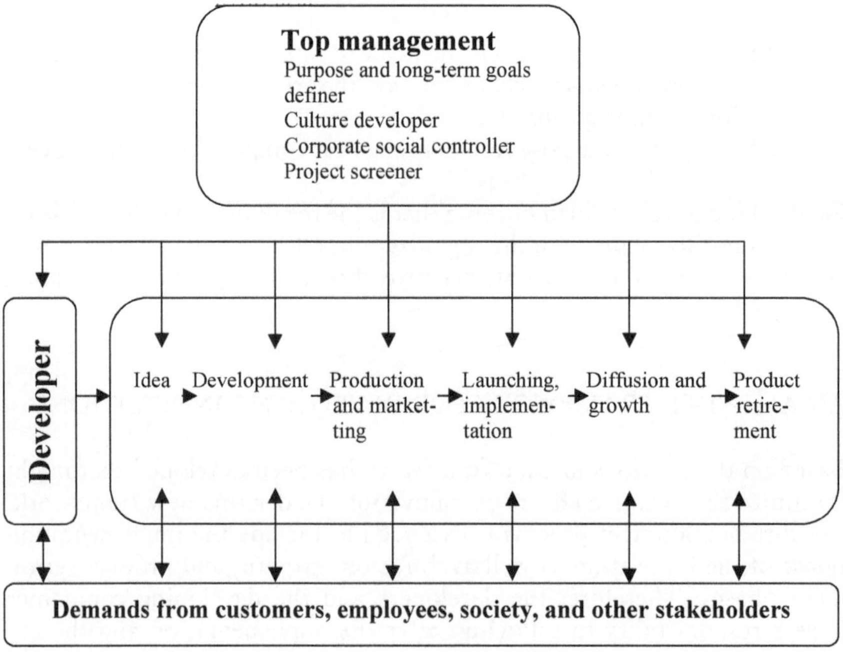

Figure 8.1 An outline of an innovation model that minimizes the undesired and unintended effects of innovations.

REFLECTIONS AND CONCLUSION

What are the concrete implications of this framework on developers and top management? For one, the focus of future innovation models should not only be on the technological aspects or the commercial success but should also involve greater emphasis on the implementation and diffusion of innovations. The future models must address the effects of the innovations on all stakeholders, including the employees and society as a whole as. There should be less need for what Bessard and Tidd called social entrepreneurship and what Ahmed and Shepherd termed political innovation. The companies must take action in advance to eliminate negative consequences of innovations, which then begs the question: How should innovation be managed?

For *a short-term development and profit focus*, existing models seem to fulfill their role as a support for management and for manager decision making. But from the long-term perspective, the most important actions include introducing into the company a stakeholder view and developing the management function within the company. When a proper top management philosophy is in place, companies are focused more on the long term and the ethical rights of stakeholders are not being violated. As Ellsworth (2002) has indicated, a key element, when defining the basic purpose of a company, is to consider and prioritize the expectations and demands of the stakeholders, thus creating a corporate culture and climate that supports an ethical, long-term commitment to innovations. In such an environment, companies are more willing to address and correct, in the early phases, possible future negative consequences. It must also be emphasized that decentralization and involvement of employees usually generates much more positive energy. Our main argument, therefore, in this study is that future innovation models should be more connected to their context, both internally and externally, thereby minimizing undesirable consequences of innovations and resulting in a better world for all stakeholders. Further, as the models reviewed here form a core for, and are largely disseminated through, innovation management and entrepreneurial education, it is important also for educators to bring in more reflective models, such as the one proposed here, to minimize negative effects of innovation at an early stage.

REFERENCES

Abernathy, W., & Utterback, M. (1978). Patterns of industrial innovation. *Technology Review*, 80, 41–47.

Adizes, I. (1988). *Corporate lifecycles: How and why corporations grow and die and what to do about it.* Englewood Cliffs, NJ: Prentice Hall.

Ahmed, P. K., & Shepherd, C. D. (2010). *Innovation management: Context, strategies, systems and processes.* Essex: Pearson Education.

Appelbaum, E., Baily, T., Berg, P., & Kalleberg, A. L. (2000). *Manufacturing advantage: Why high-performance work systems pay off.* Ithaca, NY: Cornell University Press.

Benavente, J. M., & Lauterbach, R. (2007). *The effects of innovation on employment: Evidence from Chilean firms* (Working Paper). Santiago: Facultad de Economía y Administración.

Bessant, J., & Tidd, J. (2007). *Innovation and entrepreneurship*. Chichester: John Wiley.

Bruce, B. C. (1993). Innovation and social change. In B. C. Bruce, J. K. Peyton, & T. W. Batson (Eds.), *Network-based classrooms: Promises and realities* (pp. 9–32). New York: Cambridge University Press.

Burgelman, R. A. (1983, June). A process model of internal corporate venturing in diversified major firm. *Administrative Science Quarterly*, 223–244.

Burgelman, R. A., & Sayles, L. R. (1985). *Inside corporate innovation: Strategy, structure and managerial skills*. London: Free Press.

Bryson, A., Dale-Olsen, H., & Barth, E. (2009). *How does innovation affect worker well-being?* (Centre of Economic Performance Discussion Papers No. 0953). London: London School of Economics and Political Science.

Conway, S., & Steward, F. (2009). *Managing and shaping innovation*. Oxford: Oxford University Press.

Cooper, R. G. (2001). *Winning at new products*. New York: Perseus.

Cooper, R. G., & Kleinschmidt, E. J. (1986). An investigation into the new product process: Steps, deficiencies, impact. *Journal of Product Innovation Management*, 3, 71–85.

De Witte, H. (1999). Job insecurity and psychological wellbeing: Review of the literature and exploration of some unsolved issues. *European Journal of Work and Organizational Psychology*, 8, 155–177.

Ellsworth, R. E. (2002). *Leading with purpose: The new corporate realities*. Palo Alto, CA: Stanford University Press

Etienne, E. C. (1981). Interaction between product R&D and process technology. *Research Management*, 24, 22–27.

Freeman, R., & Kleiner, M. (2000). Who benefits most from employee involvement: Firms or workers. *American Economic Review*, 90, 219–223.

Foster, R. (1986). *Innovation: The attacker's advantage*. New York: Summit Books.

Gardiner, P., & Rothwell, R. (1985). Tough customers: Good designers. *Design Studies*, 6, 7–17.

Greiner, L. E. (1972). Evaluation and revolution as organizations grow. *Harvard Business Review*, 50, 37–46.

Harrison, R., Jaumandreu, J., Mairesse, J., & Peters, B. (2008). *Does innovation stimulate employment? A firm-level analysis using comparable micro-data from four European countries* (Working Paper No. 14216). Cambridge, MA: National Bureau of Economic Research.

Harrisson, D., & Legendre, C. (2003). Technological innovations, organizational change and workplace accident prevention. *Safety Science*, 41, 319–338.

Johne, F. A., & Snelson, P. A. (1988). Success factors in product innovation: A selective review of the literature. *Journal of Product Innovation Management*, 5, 114–128.

Kalmi, P., & Kauhanen, A. (2005). *Workplace innovations and employee outcomes: Evidence from a representative employee survey* (Working Paper No. W-395). Helsinki: Helsinki School of Economics.

Kanter, R. (1985). Supporting innovation and venture development in established companies. *Journal of Business Venturing*, 1, 47–60.

Kotter, J. P., & Heskett, J. L. (1992). *Corporate culture and performance*. New York: Free Press.

Lachenmaier, S., & Rottman, H. (2007). Effects of innovation on employment: A dynamic panel analysis. *International Journal of Industrial Organization*, 29, 210–220.

Langrish, J., Gibbons, M., Evans, W. G., & Jevons, F. R. (1972). *Wealth from knowledge: Studies of innovation in industry*. London: Macmillan.

Levitt, T. (1965). Exploit the product life cycle. *Harvard Business Review*, 43, 81–94.

Mair, J., & Marti, I. (2006). Social entrepreneurship research: A source of explanation, prediction and delight. *Journal of World Business*, 41, 36–44.

Miller, D., & Friesen, P. H. (1983). Successful and unsuccessful phases of the corporate life cycle. *Organization Studies*, 4, 339–356.

Moore, G. A. (2001). *Crossing the chasm: Marketing and selling technology products to mainstream customers*. Oxford: Capstone.

Moore, W. L., & Tushman, M. L. (1982). Managing innovation over the product life cycle. In M. L. Tushman & W. L. Moore (Eds.), *Readings in the management of innovation* (pp. 131–150). Boston: Pitman.

Peredo, A. M., & McLean, M. (2006). Social entrepreneurship: A critical review of the concept. *Journal of World Business*, 41, 56–65.

Peters, B. (2005). *Employment effects of different innovation activities: New micro econometric evidence*. Mannheim: Centre for European Economic Research, Physica-Verlag.

Pianta, M. (2006). Innovation and employment. In J. Fagerberg, D. Mowery, & R. Nelson (Eds.), *Handbook of innovation* (pp. 568–598). Oxford: Oxford University Press.

Porter, M. E., & Van der Linde, C. (1995, September–October). Green and competitive: Ending the stalemate. *Harvard Business Review*, 126–134.

Ramsay, H., Scholatios, D., & Harley, B. (2000). Employees and high-performance work systems: Testing inside the black box. *British Journal of Industrial Relations*, 38, 501–531.

Robertson, T. S. (1967). The process of innovation and the diffusion of innovation. *Journal of Marketing*, 31, 14–19.

Rogers, E. M. (1995). *Diffusion of innovations*. New York: Free Press.

Rothwell, R. (1972). *Factors for success in industrial innovations, from project SAPPHO: A comparative study of success and failure in industrial innovations*. Sussex: Brighten.

Schumpeter, J. A. (1959). *Capitalism, socialism and democracy*. London: George Allen & Unwin.

Schon, D. (1967). *Technology and change*. London: Pergamon Press.

Smith, H. J. (2003). The shareholders vs. stakeholders' debate. *MIT Sloan Management Review*, 44, 85–90.

Stayer, D. L., Drews, F. A., & Crough, D. J. (2006). A comparison of cell phone driver and drunk driver. *Human Factors: The Journal of the Human Factors and Ergonomics Society*, 48, 381–391.

Tushman, M., & Anderson, P. (1986). Technological discontinuities and organizational environments. *Administrative Science Quarterly*, 31, 439–465.

Urban, G. L., & Hauser, J. R. (1980). *Design and marketing of new products*. London: Prentice-Hall.

Ytterhus, B. E., & Synnestvedt, T. (1995). *The process of greening: Results from the Nordic business environmental barometer* (Working Paper No. 32). Sandvika: Norwegian School of Management.

Zahra, S. A., Gedajlovic, E., Neubaum, D. O., & Shulman, J. M. (2009). A typology of social entrepreneurs: Motives, search processes and ethical challenges. *Journal of Business Venturing*, 24, 519–532.

Zimmermann, K. F. (1991). The employment consequences of technological advances, demand and labor costs in 16 German industries. *Empirical Economics*, 16, 253–266.

Zirger, B. J., & Maidique, M. A. (1990). A model of new product development: An empirical test. *Management Science*, 36, 867–884.

Zwick, T. (2002). Employee resistance against innovations. *International Journal of Manpower*, 23, 542–552.

9 From Autonomous Craftsmen to Compliant Resources

Implications for Undesirable Consequences of Innovation

Beata Segercrantz

This chapter examines software innovation and what implications the restructurings that increase formalized ways of organizing software innovation have for information and communication technology (ICT) experts. The study draws on interviews conducted with ICT experts in Finnish small and middle size ICT firms. The focus is on the interplay between formalization of ways of organizing innovation and ICT experts' transforming possibilities to negotiate agency (for example, exercise control over their work). More specifically, the chapter explores the following questions: When ways of organizing software innovation become more formalized, how are the ICT experts involved positioned? What are the implications of the interplay between formalization and ICT experts' agency and possibilities to address unintended and undesirable consequences of the products they develop? Findings suggest that when organizing becomes more formalized, ICT experts often become positioned increasingly as 'compliant resources' rather than as 'autonomous craftsmen.' This limits their possibilities to negotiate agency and consider unintended and undesirable consequences of innovation.

There are several reasons to explore the issues raised in this chapter. The context of this study is located within the ICT sector, including the software industry. This sector is often considered, in many national contexts, as an important (potential) driving force of economic growth and thus also (in)directly of social change. Whether one agrees with this view or not, it is difficult to deny that the ICT sector has gained much attention and a central position on various societal levels, such as policy making and national research and development (R&D) expenditure. As a result of the technological emphasis and development during the past decades, ICT and software, in particular, are today ubiquitous in nearly all aspects of developed capitalist economies (Case & Pineiro, 2009). As many authors have concluded, carrying out even simple everyday tasks would be difficult or even impossible without ICT. Consequently, various direct and indirect unintended and/or undesirable consequences of software innovation may affect many.

Considering the importance of ICT innovation in contemporary policy making, on the global market and not the least in everyday life, it seems important to explore within the field of innovation studies the unintended and undesirable consequences of innovation, for example, in terms of various actors' possibilities to reduce them. There are of course many potential social actors who could engage in such processes; this chapter will focus on ICT experts involved in innovation and new product development (NPD) by providing empirical illustrations and analyses. The interest is in these experts' lived experiences of the very mundane local conditions of software innovation; conditions that they engage in through their daily work. This is because such conditions could be important for considering and identifying possible future unintended and undesirable consequences of innovation (see chapters 6, 10, and 11 in this volume).

While taking an interest in these mundane local conditions of software innovation, particular attention is given to efforts of increasing formalization and standardization of daily innovation practices. This is to provide a glimpse into the possibilities and/or limitations formalization may provide ICT experts with regarding considering the consequences of the products they develop. In this chapter, formalization is defined as the efforts to move away from improvised ways of organizing and instead toward shared, stricter, standardized, and/or less flexible ways of organizing.

The chapter is structured as follows. First, software innovation and tension related to organizing such innovation are discussed in terms of project management. Following this, discourse analysis is discussed as a useful methodology for studying processes that constitute software innovation. The chapter then moves on to presenting and analyzing an empirical case. The analysis explores how formalization is imposed through ways of organizing innovation as well as the implications for ICT experts. Finally, the chapter engages in a discussion regarding the interplay between formalization and the ICT experts' agency and possibilities to address unintended and undesirable consequences of the products they develop.

WAYS OF ORGANIZING SOFTWARE INNOVATION

Ways of organizing innovation, that is, mainly various processes and practices, have been studied and theorized extensively. Pavitt (2006) argues that Schumpeter's analyses of innovation represent perhaps the first influences in this field of research. However, as he stresses, the study of innovation processes has expanded during the last decades far beyond the first attempts to theorize innovation processes. Over time, various established approaches to innovation and new product development (NPD) processes have emerged. These approaches typically investigate how innovation and NPD practices are organized and managed and how innovation processes can be improved; how knowledge is produced, transformed into different types of innovations, and matched to market needs and demands (Pavitt, 2006). From having argued that innovation and NPD involve processes that can be carefully organized, managed, and formalized (see, for example, Cooper, 1990), scholars are increasingly recognizing that innovation practices are complex, political, and unpredictable phenomena 'contingent on sector, firm, and technology field' (Pavitt, 2006, p. 109).

Pineiro (2003) argues that mainstream studies of software programming (and perhaps also technological innovation in general) typically assume that innovation is an objective instrumental activity that can and should be optimized and made efficient to overcome the problems arising from the erratic nature of programming and to achieve certain results successfully. Moreover, viewing software innovation from this perspective involves efforts to increase ordering, control, and formalization in work processes. Unsuccessful software innovation is thus typically viewed as a result of insufficient project management or techniques (Pineiro, 2003).

Unsurprisingly then, there are numerous management models (e.g., project management) that focus on exploring and finding the key to successful techniques of organizing innovation; some of them are examined in chapter 8 of this volume. Taking project management as an example of a management model, one can see that over time, project-based work has increasingly been viewed both by scholars and practitioners as an efficient alternative to standardization and bureaucracy (Clegg & Courpasson, 2004) in order to deal with continuous change and unpredictable work in the context of uniqueness and temporariness (Cicmil, Hodgson, Lindgren, & Packendorff, 2009), to bring about creativity and innovativeness, and to meet many of the requirements of innovation.

In addition, project management has increasingly been seen as a particular methodology, tool, or technique for accomplishing objectives efficiently and on time, but also as a way of organizing and structuring. Such ways of organizing work emphasize the flexible structures of the project while simultaneously imposing more permanent structures to 'professionalize' project management and to construct managerial control (Cicmil et al., 2009; Clegg & Courpasson, 2004). Herein lies a paradox particularly relevant for innovation: Project work, as most other formalized innovation processes, is both seen as an alternative to standardization and bureaucracy and as a form of work that draws on sophisticated methods that invite standardization and specialization (Cicmil et al., 2009).

The emphasis on optimizing and controlling software innovation through increased specialization as a response to meeting continuous new challenges seems to invoke a reified understanding of this work. More specifically, work is viewed as a controlled activity that is only carried out to reach certain outcomes, for example, the commercialization of a new software product (Pineiro, 2003).

To sum up, both traditional innovation models and models claiming to promote less bureaucratic techniques seem to invoke formalization (although of different kinds) and an instrumental, unambiguous view of ways of organizing innovation. Whereas these conceptualizations of innovation may produce methodologies that improve productivity, they also typically imply that the heterogeneous unfolding processes of innovation are ignored, including broader consequences on different actors and society. The actual 'doing of innovation' (Pineiro, 2003), the microprocesses and becoming of innovation, are ignored, as are the tensions and ambiguities involved. This chapter examines these ignored microprocesses to understand the implications for ICT experts involved in software innovation undergoing formalization. The chapter is particularly focused on their possibilities to negotiate agency and to consider consequences of the products or services they are developing.

THE DISCURSIVE AND MUTUAL CONSTRUCTION OF SOFTWARE INNOVATION AND SUBJECT POSITIONS

This study draws on discourse analysis as this methodology offers an opportunity to analyze *processes* of social construction (Berger & Luckmann, 1966; Gergen, 1985; Phillips & Hardy, 2002) that constitute and interconnect innovation, organizing, subject positions, and agency (Davies & Harré, 1990). However, it must be emphasized that discourse analysis can and has been adopted in a variety of different ways and at differing levels (Alvesson & Karreman, 2000). Therefore, how discourse analysis will be adopted in this study is explained in detail next.

The adopted discursive approach to social reality places focus on everyday interaction and social construction between people (Berger & Luckmann, 1966; Burr, 1995); how language constitutes social reality rather than reflects it (Jokinen & Juhila, 1999; Phillips & Hardy, 2002). In and through interaction and language, people are seen to be provided with certain categories of social reality, such as innovation, while they simultaneously also participate in maintaining or contesting meanings attributed to these categories. Here, discourse can be seen as practices of talking and writing, which brings objects or categories into being. For example, the dominant discourse of innovation constructs technological innovation as a prerequisite of economic growth. Moreover, as Phillips and Hardy (2002, p. 3) argue 'social reality is produced and made real through discourses, and social interactions cannot be fully understood without reference to the discourses that give them meaning.'

When exploring the discursive construction of software innovation, as any social reality, we can expect to come across a complex web of interrelated discourses; a variety of heterogeneous discourses always 'come together' in specific contexts. Furthermore, we can assume that some local conditions are constituted by a wide range of discourses, whereas other conditions provide actors involved with a very restricted number of discourses to draw on. In this context, dominant discourses are likely to afford some individuals, groups, or managers more agency regarding organizing innovation, whereas consequently other voices are marginalized or silenced (Grant, Michelson, Oswick, & Wailes, 2005; Potter & Wetherell, 1987). Social actors involved in software innovation are, in other words, in varying degrees both users and manipulators of discourse dependent on prevailing power relations and their positioning. In this chapter I am, in particular, interested in how ICT experts are positioned and thus users/manipulators of discourse through power relations.

The discursive production of subject positions can be explored, for instance, by an analysis of how a social actor's actions in specific local conditions are interrelated to how she or he is addressed and positioned in discourse (Davies & Harré, 1990). The participation in discursive practices involves positioning the self in relation to different categories, for example, as a white male software expert, and thus positioning also involves a process through which the self is differentiated from the other (Davies & Harré, 1990).

In our daily enactments, we are constantly addressed by a variety of different discourses. This implies that depending on time and space, we may (or may not) take up a variety of different subject positions, and consequently the processes of positioning involve a production of a diversity of selves. One could say that we live in a flow of changing positions, some of which we may take up 'temporarily,' whereas others become more 'permanent' ones; some of which we find unproblematic, whereas others are experienced as contradictory; some of which we accept and/or identify with, others that we resist and/or disidentify with; some of which we are relatively aware of, others that we draw on unintentionally. Nevertheless, despite the diversity of selves and potential contradictions, people typically understand themselves as the sum of various subject positions, and they tend to view themselves as historically continuous and unitary. However, the suggestion here is that these practices are ongoing, political, and related to power relations, and so *who one is* is constantly in a state of flux, and any definition depends on those discourses available to and taken up by a person and others within a specific practice (Gergen, 1999).

Therefore, processes of positioning have more or less radical implications for ICT experts' agency. Agency is theorized as interconnected to the social actors' differing 'access' to discourse and varying possibilities to 'manipulate' them (Burr, 1995). In that sense, the study takes neither a Foucauldian deterministic nor a voluntarist view on discourse, subject positions, and agency. Rather the study attempts to show *how* and *when* discourses and subject positions mutually emerge (Bergström & Knights, 2006). The discursive view of positioning taken here then suggests that social actors take up specific subject positions within discourse, and that this is achieved through the interaction between discourse and human agency (Bergström & Knights, 2006).

THE INTERVIEW MATERIAL: STUDYING DISCOURSE

Longitudinal interview material gathered for a larger study[1] during four different phases between 2001 and 2005 has been drawn on in this chapter. The study consists of 81 semi-structured interviews conducted in Finnish small and middle size ICT companies. Of these 81 interviews, a discourse analysis was done on 34 interviews with 22 ICT experts, which this chapter draws on. The interviews conducted covered issues around NPD, innovation, and organizational restructurings.

Common to all the experts interviewed was that they worked with software development or innovation.[2] I have decided to use the term 'expert' because all the interviewees were experts in something related to software development. However, it is not only the work description that is similar to all interviewees. The way in which the interviewees were selected in the study through a particular snowball sampling technique also meant that, for example, gender, age, ethnicity, nationality, and education were not used as selection criteria of the interviewees. This contributed to an interestingly homogeneous group of interviewees in terms of these social categories. The typical interviewee was a white, Finnish, 30-year-old male with a completed or partially completed master's degree in subjects related to technology and/or engineering. This is important to keep in mind while reading this chapter, as when the interviewees talk, they take up certain subject positions (that of, for instance, craftsmen) that shape their perspectives on reality and make the interview material gendered.

The overall methodology and also the analysis put emphasis on the empirical data and issues arising from it. Therefore, the study avoided formulating specific presumptions at the outset, although the aim was to facilitate a critical account following the analysis and not only to illustrate the becoming of particular discursive effects. An important beneficial implication of this effort is the possibility to give voice to a broader array of accounts, perhaps previously marginalized ones, and address some consequences of software innovation. Therefore, the intention was to be as open as possible to different constructions of innovation in the analysis, and then after the analysis move toward a more critical account.

The analysis began, as most discourse analyses, by readings and rereadings of the material.[3] The analysis and interpretations occurred on the basis of 'structured' readings and writing, through which I began to see that the interviewees often talked about certain themes, even when not explicitly asked about them. This insight allowed me to explore how certain meanings were both explicitly and implicitly constructed. First, I analyzed texts in which the interviewees explicitly constructed meanings around innovation and ways of organizing by paying attention to how they drew on certain discourses. Second, the analysis also examined how the interviewees simultaneously and implicitly constructed and positioned themselves as persons, even when they were not explicitly asked to do so.

AN ILLUSTRATIVE EXAMPLE: FORMALIZATION, POSITIONING, AND AGENCY

To highlight some crucial overall findings from this discourse analysis, I will provide an illustrative example in this section. The discourse analysis was done on four groups of interviewees, each group representing one specific firm. For the illustrative example in this section, one of these four organizations (hereafter given the pseudonym Sissela) was chosen, and it included accounts of all five interviewees from this organization: Frank, Ben, Cliff, Neil, and Richard. Although I only use quotations from one organization, I wish to emphasize that the illustrative example illuminates the overall findings from the discourse analysis of 34 interviews in 4 organizations.

Setting: Moving Toward Formalization at Sissela

Sissela, a company of nearly 100 employees, develops mobile and Internet software for operators and network equipment manufacturers. Sissela was founded in the early 2000s, but it has a background of acquisitions, periods of rapid growth, and a downsizing process that stretch some years back from that. Over these years, the ownership, name of the company, and number of employees have changed several times. Consequently, Sissela's innovation and NPD practices have taken form in a context characterized by ongoing restructurings, as have many other ICT companies' practices, too.

When accounting for ways of organizing software development, the interviewees focused in particular on a period of rapid growth, downsizing, and adaptation to subcontracting. Therefore, this section explores how formalization was increasingly imposed over time as these three restructurings unfolded.

Two of the five Sissela interviewees had experienced the periods of rapid growth and downsizing, which seemed to differ significantly from the period of subcontracting when the other three interviewees were recruited to Sissela. During periods of growth and downsizing, ways of organizing had typically been autonomous and improvised, as two ICT experts explained:

> What is mainly emphasized is the use of heuristics and the designer's own visions. . . . This is not like the car industry, which already has well-established practices according to which products are produced in specific ways.

> It [the product development process] became faster and more flexible. Probably because the organization was smaller, the communication is in a way easier. When you sit next to each other, then you lose, to some extent a, a, what should I call it, a larger process. Or it develops more easily into just working, you just do things and not according to a process, the documentation is not necessarily done that well. It's about having to get lots of things done and you don't have time to think about the process in itself.

As the interviewees describe here, both during rapid growth (first extract) and during a downsizing process (the second extract) the ICT experts felt that the ways of organizing were relatively improvised and informal.

After having survived a heavy downsizing process around the burst of the IT-bubble in the early 2000s Sissela managed to sell a long-term project to another Finnish ICT company. This subcontracting project implied that ways of organizing were restructured significantly. Most activities seemed to be dictated by specific ways of organizing, that is, a formalized structure and planning, chains of checks, and networks of relations. For the product development experts, this meant a strong focus on narrowly defined work tasks and compliance to prespecified plans. One of the interviewees explained:

> We have these plans, according to which we live. And schedules have been made early on in the project, which of course change and are drawn out and also become shorter. But in that way quite a lot has been done in advance. That is, certain stages have been given, which define what has to be ready by a specific point in time.

As seen here, the move from the period of downsizing to the subcontracting project implied that the daily practices of innovation and ways of organizing changed. The account forms a stark contrast to the interviewees' earlier experiences of more improvised organizing and broader work descriptions. Now the interviewees daily work tasks took shape in a context of more bureaucratic and 'formal' ways of organizing. The daily work and organizing were controlled by complex formalized structures. However, working under the pressure of these structures were not always easy:

> What we actually have done inversely—for example, I have already done and run half of the tasks before having finished the plan. But it is really a consequence of me having to work according to too tight timeframes. That task was given to me too late. There wouldn't have been any point in writing it [the documentation for the task] then. Now I have to do it reversely and work in a tremendous hurry, at the same time as I'm implementing, I'm writing the plan.

As the quote suggests, the ICT experts were committed to comply with project goals and guidelines, although sometimes even in reverse. Hence, rather than focusing on producing a product of good quality, project goals and guidelines seemed to primarily guide innovation. Moreover, the formalized ways of organizing seemed to have implications for the objectives and (quality of) products.

What is also of interest here is how subject positions are constructed, offered, changed, and taken up in action within the meanings ascribed to the increasingly formalized ways of organizing innovation and NPD: What are the implications of the interplay between formalization and positioning for the actors' agency? Next this question will be addressed.

Positioning and Agency: From Autonomous Craftsmen Toward Being Compliant Resources

In this section, how the five ICT experts at Sissela are positioned in relation to increasingly formalized ways of organizing software innovation are examined. The two ICT experts, who had experienced rapid growth at Sissela, described organizing during this period as relatively improvised and informal. Within their accounts one can see how they drew on particular subject positions offered in the context of the particular prevailing ways of organizing.

In relation to the improvised and informal ways of organizing, the interviewees constructed themselves foremost as autonomous workers. This was accomplished through discursive practices where the interviewees were provided with the possibility of controlling how projects unfolded and developing a holistic understanding of it. As we already heard one expert say in the previous section, 'the designer's own visions' were emphasized, thus implying that the experts were seen as autonomous actors or, as his coworker claimed, during the period of rapid growth he was enthusiastic, committed, and in control of his own work. More specifically, the ICT experts highlighted their skills above hierarchical positions by addressing themselves in terms of autonomous 'craftsmen' and identifying with this subject position. One could argue that the Sissela interviewees seemed to have relatively good possibilities to negotiate voice and agency regarding, for example, ways of organizing.

However, the period of rapid growth was soon followed by a burst of the IT-bubble and a heavy downsizing process. This created divisions and tensions between and within various social groups within the organization. Taking up subject positions within the organizational network of social relations seemed to require ongoing processes of positioning. The two interviewees who were working for the company at this stage explained that the ICT experts were in the initial phase of the downsizing process split into two groups of 'doers' and 'non-doers':

> I, myself, had work to do all the time, which helped me pull through, so that I could focus on it and in a way push away other people a little but it was some sort of hiding.

Here, the interviewee is positioned as a doer. Still the interviewees were now provided with fewer possibilities to exercise control over the projects they were involved in. An important reason was that the sales department was prioritized and given better possibilities to voice their preferred interests as a response to the financial problems, thus altering the ICT experts' previous autonomous subject position.

> The sales department sells something, something to the customer with an unrealistic timeframe . . . and then it starts off in a dreadful hurry or from something strange that nobody will benefit from.

The ICT experts were, in other words, increasingly experiencing subordination in relation to others in the organization, such as the employees in the sales department, and in that sense, possibilities to negotiate agency became restricted. Yet, despite frustration with the tensions and subordination, the interviewees were prepared to carry out obligations attributed to the subject position offered in discourse. This was done by making sense of the local conditions that the company operated within in a very specific way ('a small company' without 'enough resources'). By making sense of the organization in these terms, the ICT experts continued to carry out their work despite the undesirable subject positions. This in turn worked as discursive practices in and through which the interviewee positioning and agency shifted (Alvesson & Willmott, 2002). Moreover, as a group, the ICT experts were offered the subject positions of 'doers' (whereas most non-doers were dismissed) and as 'multitasking-actors' but also that of 'subordinates' and 'obedient' in contrast to being addressed as autonomous experts.

After the downsizing process, Sissela only employed approximately 20 employees. Developing software by drawing on previous ways of organizing the firm managed to increasingly sell products and services, including a long-term subcontracting project to another Finnish ICT firm. However, this involved moving an innovation team to the other ICT firm's premises. Once having moved there, the employees of Sissela had to begin to adapt to their customer's ways of organizing. This is the project the other three interviewees were employed for by Sissela.

The new subcontracting project was described as predictable, planned, and structured. This understanding of the project had very specific implications for how work was carried out and for how the social actors took up subject positions. More specifically, an overall plan, developed by the customer of the subcontracting project, specified rules and guidelines for the NPD project. Hence, the overall plan acted as a discursive resource that invoked how to act in various settings and consequently had implications for various social actors' positioning. Two ICT experts explained:

> . . . internally our customer has created an overall project plan, according to which parts then are just handed out "to you," "to you," and " to you."

> There is quite a chain of checks. You always have to ask so many in higher quarters "Can I do this?"

As we see here, the ICT experts were addressed as passive receivers of certain tasks, that is, defined on the basis of skills and addressed as experts on specific parts of the product. Autonomous work was again restricted through 'chains of checks' and control. Moreover, to carry out the obligations attributed to the expert subject position, the interviewees were expected to continuously seek approval of their plans and of their work from other actors. This became an important arena for negotiating subject positions and power relations. The interaction implied a constant reinforcement of the established 'productive expert' subject positions offered through the overall plan as the ICT experts sought approval for their work to make sure they fitted into the whole. Perhaps not surprisingly, many developed a rather instrumental view of themselves.

However, despite some struggles, the formalized ways of organizing were typically embraced by the interviewees:

> Having these tools makes it easier in the sense that there is a certain practice that everyone is committed to.

> Of course it's a good thing that things are [short break], kept together in a controlled way.

Although the interviewees embraced the highly formalized ways of organizing, as seen in their quotes, many of the interviewees expressed discomfort with the subject positions offered within this context, as one interviewee said:

> When you look at it from a far distance, it's been a good, good experience, but not my thing in the end. . . . My wish is that I could return to some kind of handicraft one day.

As this section has shown, the adaptation to subcontracting involved adopting and complying with highly formalized organizing drawing on a prespecified plan that could be said to have acted as a discursive resource (Tsoukas & Chia, 2002). The processes triggered by this discursive resource implied that the interviewees were addressed as productive resources, as relatively passive receivers of tasks, and as experts with specific skills. This in turn implied that the interviewees became subordinate to the plan and, in various ways, disidentified with their subject positions by constructing a desire for more autonomy or agency to develop as professionals or 'craftsmen.' For some, this involved a wish for reviving lost subject positions, and for others this meant looking into new opportunities in the future.

DISCUSSION

The analysis of mundane, everyday innovation practices including implications for subject positions shows how the interviewees' subject positions shifted so that from having been addressed as 'autonomous craftsmen' during rapid growth and multitasking to subordinate actors during downsizing, the ICT experts were increasingly positioned in terms of productive resources, passive receivers of tasks, and skillful experts in the subcontracting project. This shift toward fewer opportunities to negotiate agency also seemed to imply a shift from identification with the subject positions toward disidentification; the ICT experts silently dissociated themselves from the formalized ways of organizing as agency was restricted, although they simultaneously accept and legitimize the formalized ways of organizing as efficient.

Drawing on these findings, this chapter opens up a discussion about unfolding innovation practices in terms of interconnections between increasing formalization, agency, (silent) disidentification, and compliance. Although the organizations participating in this study were different in many respects, they all seemed to share an increasing desire for greater order and formalization of ways of organizing, which in turn contributed to fewer possibilities for the ICT experts to negotiate agency and to 'silent' disidentification. Thus, the analysis highlights how innovation may work through the reproduction of power relations, which social actors, such as ICT experts, paradoxically seek to distance themselves from (Fleming & Spicer, 2003).

Moreover, the empirical illustrations show how the interviewees' actions and possibilities to negotiate voice were dependent on how they were addressed and positioned in relation to prevailing discourses of innovation and organizing, which provided them with certain structures of rights and obligations (Davies & Harré, 1990). In that regard, the study has touched on issues of agency.

How one understands agency is of course dependent on perspective and on how one views discourse (Burr, 1995). In this study, positioning, discourse, and agency are seen as mutually and dynamically produced and therefore studied as emerging in the interaction between different actors and power relations (rather than turning the focus on the individual) (Bergström & Knights, 2006). Drawing on this view emphasizing agency as intertwined with positioning and discourse, agency can be seen as potentially taking different forms in different contexts at different points in time. More specifically, particular discursive practices, as show in the illustrations from Sissela, may contribute to different forms of agency for different social actors, which in turn may change over time. Agency is hence interconnected to the social actors' differing 'access' to discourse and varying possibilities to 'manipulate' them (Burr, 1995).

This view on agency therefore suggests that change is at least a possibility if prevailing discourses are challenged or resisted and alternative subject positions and agency are made available to social actors although such trajectories were not obvious in the organizations studied. However, I wish to emphasize the potential opportunities of change as this is important if we are to address unintended and undesirable consequences of innovation.

CONCLUSIONS AND IMPLICATIONS

This chapter has sought to explore the implications that increased formalized ways of organizing innovation have for ICT experts. The focus has been on the interplay between formalization of ways of organizing innovation and ICT experts' transforming possibilities to negotiate agency through subject positioning. The chapter poses the questions: When ways of organizing software innovation become more formalized, how are ICT experts involved positioned? What are the implications of the interplay between formalization and positioning for the actors' agency and possibilities to address unintended and undesirable consequences of the products they develop?

So far the chapter has mainly addressed the first research question. I have shown how increasing formalization in software innovation is intertwined with power relations, which discipline social actors and restrict possibilities to negotiate agency and voice. In the empirical illustration we could see that when formalization in the daily work practices increased, the software experts interviewed engaged increasingly in processes of disidentification (Fleming & Spicer, 2003) through ongoing processes of positioning. Over time the formalization implied that the ICT experts (silently) began to dissociate themselves from the restricting local conditions that provided them with little choice within discourse. However, this is not to say that the ICT experts were passive receivers of discourse. The analysis shows how they simultaneously both were produced in discourse and actively participated in the reproduction of discourses and power relations they sought to distance themselves from (Fleming & Spicer, 2003).

The argument in this chapter is that the ICT experts' disidentification with subject positions offered in and through discourse and decreasing room for choice are unintended and often undesirable consequences of innovation, and a phenomenon deserving more attention in future research. Yet it must be emphasized that disidentification is a contested phenomena that is dependent on perspective and context. For example, Fleming and Sturdy (2011) as well as Fleming and Spicer (2003) argue that both identification and disidentification may have negative and positive implications for different individuals and organizations. Thus, disidentification is not necessarily an undesirable consequence of innovation, although it often may be an unintended consequence. However, processes of disidentification and perhaps foremost decreasing agency and choice within discourse may have further implication, which the second research question touches on by asking what the implications of the interplay between formalization and positioning are for the actors' agency and possibilities to address unintended and undesirable consequences of the products they develop.

Although this chapter does not provide empirically grounded answers to this question, the findings of the study raise some crucial issues. In particular, the study draws attention to how formalization in software innovation is experienced by ICT experts as a trap of requirements that reduces agency and voice. As one interviewee claimed, the formalized local conditions meant that 'you just have to try to produce a satisfying solution on the basis of all the requirements from various directions.' When ways of organizing software innovation become more patterned, standardized, or formalized, work descriptions are likely to become more specific and often more narrow (as at Sissela). It is then also likely that ICT workers lose a more holistic perspective on the innovation processes they are involved in. As three interviewees said:

> I have lost it [the ability to know in what direction to develop products], once I used to have an understanding of it, now I have realized that I don't have the capacity anymore, I can't do it.

> But then these internal, smaller projects are divided into internal, smaller parts. And it feels like every matter has some sort of a manager there, who is responsible for smaller and smaller parts. On the other hand, it sometimes feels like there are more managers than people doing productive tasks.

> In a way I throw off the responsibility for defining what this product should be used for.

ICT experts may, for instance, lose responsibility of quality (quality becomes the concern of the quality control manager) and lose insight into what other close coworkers do. That is, the aim to produce a software product of good quality by drawing on a discourse of creativity is replaced by compliance to project goals, guidelines, and time pressures, which gives less time and space to critical and creative reflection (also noted in financial innovation, see chapter 6 of this volume). Therefore, this chapter also argues that the disidentification and the loss of responsibilities and critical reflection may, in turn, increase the risk of unintended and undesirable consequences of innovation.

Those individuals who do not comply with the imposed ways of organizing, 'the difficult craftsman,' risk being forced out from the organization or may leave 'voluntarily,' thus further reducing creativity and reflexivity. For instance, in one of the organizations studied, the project leader removed one ICT expert from the team, who he thought engage too much in questioning the innovation process:

> I decided that he had to be removed elsewhere, because with the schedules we had, his way of working would have taken too much of my energy, of supervision, that's why it was better that this other mate, who was sitting over there, who was very easy to cooperate with, who quickly understood the core things . . . it was easier to argue for my decisions with him . . . I had to make some radical changes, and make a decision about who could contribute to a top team that can achieve things. I had to get people to believe in this thing, that is, if we do it like this, it [the product] will become successful; we will make it in time, with the resources allocated to us.

Stated more broadly, when managerialist views are intensified and time pressures and profitability override most other organizational concerns, reflexivity decreases and societal consequences of the management models and innovation that stretch beyond the firm's immediate interest easily become suppressed, forgotten, or underestimated (Sturdy & Grey, 2003; see also chapters 2 and 5 in this volume).

On the other hand, it is possible that disidentification could become a source for critical reflexivity. When disidentifying with subject positions made available in discourse, experts may also engage in more critical reflexivity than when identifying with them (Fleming & Spicer, 2003). However, this was not the case in the organizations studied. Hence, when formalization works as a limitation for negotiating agency and voice, as shown in this study, we should be concerned about the risk that ICT experts lack possibilities to address undesirable consequences of innovation. Therefore, the aim is not here to argue for efforts to increase identification and against disidentification, but rather to show how they are interconnected to formalization and agency. These are all highly important sources and phenomena if we are to address and avoid undesirable consequences of software innovation.

To conclude, formalized ways of organizing in software innovation may have unintended and undesirable consequences; first, on the level of positioning and subjectivity and, second, on a more macro- or societal level because changes on the level of positioning may have far-reaching implications through the loss of reflexivity when formalization is imposed. This calls for more critical work that interconnects power relations, subject positions, and management models to innovation and, in particular, consequences of innovation. There is a clear need for more research in this area regarding the limitations various innovation models and processes create for agency and reflexivity, in particular, as new models are continuously imposed in many ICT firms.

NOTES

1. The interview material was gathered for a doctoral thesis (Segercrantz, 2011).

2. The interviewees' titles were: interaction designers, software/IT/technical architects, senior technology officers, software engineers, senior consultants (testing), senior technology specialists, development managers, or product development managers/directors.
3. To handle the large amount of transcripts (approximately 1,000 pages), I used NVivo (a software designed to support the handling of qualitative research materials) to divide the material into smaller parts according to certain themes.

REFERENCES

Alvesson, M., & Karreman, D. (2000). Varieties of discourse: On the study of organizations through discourse analysis. *Human Relations*, 53(9), 1125–1149.

Alvesson, M. & Willmott, H. (2002). Identity regulation as organizational control: Producing the appropriate individual. *Journal of Management Studies*, 39(5), 621–644.

Berger, P. L., & Luckmann, T. (1966). *The social construction of reality.* New York: Doubleday.

Bergström, O., & Knights, D. (2006). Organizational discourse and subjectivity: Subjectification during processes of recruitment. *Human Relations*, 59(3), 351–377.

Burr, V. (1995). *An introduction to social constructionism.* London: Routledge.

Case, P., & Pineiro, E. (2009). Stop whining, start doing! Identity conflict in project managed software production. *Ephemera Theory and Politics in Organization*, 9(2), 93–112.

Cicmil, S., Hodgson, D., Lindgren, M., & Packendorff, J. (2009). Project management behind the façade. *Ephemera Theory and Politics in Organization*, 9(2), 78–92.

Clegg, S., & Courpasson, D. (2004). Political hybrids: Tocquevillean views on project organizations. *Journal of Management Studies*, 41(4), 525–547.

Cooper, R. G. (1990). Stage-gate systems: A new tool for managing new products. *Business Horizons*, 33(3), 44–54.

Davies, B., & Harré, R. (1990). Positioning: The discursive production of selves. *Journal for the Theory of Social Behaviour*, 20(1), 43–63.

Fleming, P., & Spicer, A. (2003). Working at a cynical distance: Implications for power, subjectivity and resistance. *Organization*, 10(1), 157–179.

Fleming, P., & Sturdy, A. (2011). 'Being yourself' in the electronic sweatshop: New forms of normative control. *Human relations*, 64(2), 177–200.

Gergen, K. J. (1985). The social constructionist movement in modern psychology. *American Psychologist*, 40(3), 266–275.

Gergen, K. J. (1999). *An invitation to social construction.* London: Sage.

Grant, D., Michelson, G., Oswick, C., & Wailes, N. (2005). Guest editorial: Discourse and organizational change. *Journal of Organizational Change Management*, 18(1), 6–15.

Jokinen, A., & Juhila, K. (1999). Diskurssianalyyttisen tutkimuksen kartta. In A. Jokinen, K. Juhila, & E. Suoninen (Eds.), *Diskurssianalyysi liikkeessä*. Jyväskylä: Gummerus Kirjapaino Oy.

Pavitt, K. (2006). Innovation processes. In J. Fagerberg, D. C. Mowery, & R. Nelson, (Eds.), *The Oxford handbook of innovation* (pp. 86–114). New York: Oxford University Press.

Phillips, N., & Hardy, C. (2002). *Discourse analysis: Investigating processes of social construction.* Thousand Oaks, CA: Sage.

Pineiro, E. (2003). *The aesthetics of code: On excellence in instrumental action.* Doctorate Thesis, Stockholm, Royal Institute of Technology.

Potter, J., & Wetherell, M. (1987). *Discourse and social psychology: Beyond attitudes and behavior.* London: Sage.

Segercrantz, B. (2011). *'... the walls fell down but the blokes just coded ...' Varieties of stability in software product development during organizational restructurings.* Doctoral dissertation No226, Hanken School of Economics, Helsinki. Retrieved December 23, 2011, from: https://helda.helsinki.fi/bitstream/handle/10138/26137/226_978-952-232-122-0.pdf?sequence=1

Sturdy, A., & Grey, C. (2003). Beneath and beyond organizational change management: Exploring alternatives. *Organization,* 10(4), 651–662.

Tsoukas, R., & Chia, R. (2002). On organizational becoming: Rethinking organizational change. *Organization Science*, 13(5), 567–582.

10 Organizational Innovations

An Exploratory Study of Negative Effects

Almudena Cañibano, Oihana Basilio,* and M. Paloma Sánchez*

Since 2005, 'organizational innovation' as defined by the *Oslo Manual* has become the standard and is used by the EUROSTAT-based innovation surveys. It is 'the implementation of a new organizational method in the firm's business practices, workplace organization or external relations' (Organisation for Economic Co-operation and Development, 2005, p. 51). The main aim of this chapter is to analyze two interrelated organizational innovations included in such a definition: (1) information and communication technology (ICT) introductions that deal primarily with people and the organization of work and (2) innovative human resource management (HRM) practices, and in particular, flexible work arrangements.

Many companies are today introducing these organizational innovations on the assumption that this will improve performance (Brynjolfsson & Hitt, 1995; Truss, 2001). However, if employees are negatively affected by these innovations, not only might the company results suffer, but the performance and well-being of the workforce may be adversely affected (Ramsay, Scholarios, & Harley, 2000). Thus, some implemented organizational innovations might have undesirable and unintended consequences, even if sometimes subtle or silent (Ash, Berg, & Coiera, 2004).

We argue that to address such consequences during their implementation could counteract their negative impact on employees, thus creating an incentive for companies to assess and manage them from the start. As Godin (chapter 3, this volume) notes, we have left behind 2,500 years of history where innovation was seen as negative to enter into a period where companies, regions, and countries 'race' for innovations (chapter 5 of this volume) without paying enough attention to their possible adverse impact.

Focusing on the two organizational innovations previously mentioned, we aim to explore the following questions: (1) What are their consequences on employees? (2) Do they have any negative impact? (3) Do any other variables intervene in this relationship?

These questions are investigated through a case study of the Spanish division of a major international consultancy firm, which is suitable for our analysis for two reasons. On the one hand, it has developed cutting-edge organizational innovations in Spain that are likely to be imitated by other firms (Rodriguez-Ruiz & Martinez-Lucio, 2010), and on the other hand, it represents a critical case (Goldthorpe, 1968; Gerring, 2007) for the study of organizational innovations. Our methodology is qualitative and relies mainly on interviews.

To explore the independent variables (i.e., some organizational innovations that have recently been implemented by the company), we concentrate, on the one hand, on flexible work arrangements as a specific type of innovative HRM, focusing on two practices in particular: flexible working time and teleworking. On the other hand, we look at the introduction of three new ICT tools that deal with the organization of work (innovation, communication, and administrative tasks). Additionally, we assess the consequential possible undesirable effects on employees by contemplating two dependent variables: work intensification and decreased well-being.

In summary, this chapter intends to contribute to the literature by proposing a theoretical framework, distilled from a critical case study, that should facilitate the analysis of the undesirable effects of certain organizational innovations. We believe that detecting these negative consequences may help companies anticipate them so as to allow them to better define their strategies and apply counteractive policies. In the next section, we will discuss the extant literature on these two types of organizational innovations, focusing essentially on their link to work intensity and well-being. In section 3 we briefly describe our methodology to move on to the findings and their discussion in sections 4 and 5. The chapter ends with a succinct acknowledgement of the limitations of the study and our major conclusions.

LITERATURE REVIEW

The interest in the effects of organizational innovations is not a novelty to management literature. Indeed, many researchers have studied the business benefits that ICTs generate, for example, in terms of greater process efficiencies, new strategic alternatives, better 'visibility' of work, and so on (Brynjolfsson, 1993; Gardner, Lepak, & Bartol, 2003; Oliner & Sichel, 1994). However, research on the impact of technologies on organizations and workers has been simplistic and somewhat uncritical (Gripenberg, 2005). Political rhetoric and academic research have often been characterized by technological determinism, assuming that the right technology leads to the desired results. However, the impact of technological innovations on work is complex and contingent on a broad set of sociotechnical factors, interpretations (Avgerou, Avgerou, Ciborra, & Land, 2004; Orlikowski, 2000), and even emotional aspects or 'moods' (Ciborra, 2004), and it is impossible to fully predict the behavior of the overall system (Ash et al., 2004). In fact, research has shown that the same phenomenon can lead to diverse or even contradictory observations (Liker, Haddad, & Karlin, 1999), influenced by human action and by the individual and social representation of work and technology (Vaast & Walsham, 2005). Some authors have concluded that the only consistent finding on ICT impact on organizations is that it is paradoxical, aligned with the idea that confusion, contradiction, and complexity are part of organizational life (Robey & Boudreau, 1999).

One can find similar concerns in the human resource management literature. In the last 15 years, research in this discipline has focused on establishing whether innovative HRM practices and performance or better results are related. The predominant school of thought, mostly from the U.S., defends the existence of a positive relationship that operates through employees: introducing the practices motivates individuals into becoming more committed and performing better (Appelbaum, Bailey, & Kalleberg, 2000; Huselid, 1995; Lepak & Shaw, 2008). However, their findings have been contested for either theoretical or empirical weaknesses (Fleetwood & Hesketh, 2006; Marsden & Cañibano, 2010). Studies on flexible work arrangements have focused on their positive effects on employees, the underlying principle being that these organizational innovations allow employees to combine working and personal responsibilities (Hill et al., 2008). Nevertheless, this assumption has been challenged (Fleetwood, 2007; Higgins, Duxbury, & Johnson, 2000) and, as a result, there is still much uncertainty regarding the precise nature of the effects of innovative HRM in general, and of flexible work arrangements in particular.

One of the main problems of both ICT and HRM literatures is that the actual experience of the workforce with organizational innovations is generally overlooked. Regarding innovation models in particular, Lindell (chapter 8 in this volume) notes how the possible negative consequences both for employees and society as a whole are neglected and suggests this happens because most models do not take into account the implementation and use of the innovations. Most of the literature "uses organizations as the unit of analysis, [and] largely ignores the differences in individual employees' work attitudes and performance" (Green et al., 2006). Still, most authors assume organizational innovations systematically improve employees' well-being, without contemplating unintended or undesirable consequences of such innovations. Indeed, as noted by Pfeffer (2010), management literature tends to look at dependent variables that are indicators of firm performance, in detriment of other variables that are, nevertheless, crucial to social sustainability. Work intensification (Green, 2004a) and employee health and well-being (Pfeffer, 2010) are two such variables.

In the following paragraphs, we will review the literature that has analyzed the connections between organizational innovations and the negative effects on employees.

Work Intensification

Work intensity is defined as 'the pace of work and proportion of working hours spent in work activity' (White, McGovern, Mills, & Smeaton, 2003,

p. 179). It can therefore manifest itself as increased work time (extensive effort or hours spent on work activity), greater effort (intensive effort or pace and intensity of work), or a combination of both (Green, 2001).

During the past decade, some researchers have struck up a debate around work intensification, trying to establish to what extent it is a reality and its causes. Information technologies (Black & Lynch, 2001; Green & McIntosh, 2001) and innovative work systems (Gallie, 2005; Ramsay et al., 2000) are regarded as some of the most plausible options.

First, these organizational innovations contribute to making the working day 'boundaryless' (Harris, 2003). Laptop computers, mobile phones, and the Internet enable workers to take duties away with them, opening 'the door to more intensive work on trains, planes and at home' (Green, 2004a, p. 716), and enhancing the productivity potential of work done away from the office. They allow constant 'connectivity' and make work round-the-clock (Tarafdar, Tu, Ragu-Nathan, & Ragu-Nathan, 2007), introducing work into environments that were previously kept personal. In other words, ICTs enable employees to deliver work from 'anywhere,' filling the gaps during both work and nonwork time and reducing the 'porosity of the working day' (Green, 2004a, p. 737). With regards to flexible work arrangements, such as teleworking (or working from home), Hyman and Summers (2004) signal the danger of employees being unable to leave work, when work and home spaces are merged. Russell and colleagues (2009) find that people involved in working from home experience increased working hours. Harris's (2003) investigation corroborates these results, and her case study shows that 85% of the employees reported working longer hours. Besides this, when working from home, people tend to concentrate more and take shorter breaks (Daniels, Lamond, & Standen, 2001), increasing both work time and effort. Flexible time has also been found to have an effect on work intensity. White and colleagues (2003) show that male workers in the U.K. use their scheduling discretion to work longer hours. When work should start and finish is blurred.

Second, organizational innovations have been found to increase the pace of work. Rendering work more flexible implies passing competitive pressure onto employees, asking them to undertake more tasks, and to work faster, in other words, to work more intensely (Burchell, Lapido, & Wilkinson, 2002). This issue is closely linked to the previously described constant connectivity: with information flowing so rapidly, it seems that new work comes in faster than before, and new schedules are imposed on employees (Tolsby, 2000). Technology may duplicate work and render it more difficult and messy (Gripenberg, 2004) and, when combined with flexibility, it makes multitasking more likely, as employees have to get used to working on different things at the same time and being interrupted (Ash et al., 2004).

Third, organizational innovations oblige employees to adapt to new working conditions. To keep updated and get used to new technologies, professionals need to learn new skills (Bayo-Moriones, Billón, & Lera-López,

2008; Gardner et al., 2003), increasing the number of tasks they need to accomplish. Moreover, the more innovations, the more time they need to adapt and train. The current increased pace of change and renewal accentuates these effects (Green, 2004a), and although ICTs have a transformational impact on work, this impact 'may lead Human Resource (HR) professionals to create new innovative practices' (Gardner et al., 2003, p. 174). Likewise, organizations implementing innovative work practices tend to develop more technological innovations (De Saá-Pérez & Díaz-Díaz, 2010; Hollenstein, 2004; Tan & Nasurdin, 2010). All in all, these innovations are not only intertwined but seem to trigger one another, increasingly requiring employees to adapt and cope.

In sum, we expect organizational innovation in our study (ICTs and flexible work arrangements) to have a positive impact on work intensity.

Health and Well-Being

Despite economic progress, the decline in heavy physical labor and the legal achievements in terms of working conditions, employee well-being is found to have decreased in Europe since the 1970s (Green, 2006). Regardless of these findings, little research effort has gone into analyzing whether organizational innovations and changes affect employee well-being. In particular, Carter and colleagues (2010, p. 3) point out that regardless of 'extensive technological and organizational innovation in non-customer interactive white-collar work (. . .) it remains practically virgin territory as far as inquiry into occupational health is concerned.'

The term well-being is tied up with health. Indeed, the most widely agreed definition of health is 'a state of complete physical, mental and social well-being and not merely the absence of disease or infirmity' (World Health Organization, 1948). Bircher (2005, p. 336) adds complexity to this definition by considering health as a 'dynamic state of well-being characterized by a physical, social and mental potential, which satisfies the demands of life commensurate with age, culture and personal responsibility.' Both health and well-being are multidimensional notions that are not only affected by biological factors but also by aspects such as personality, support, public economy, or working conditions (Akerlind & Schunder, 2007). The study of the latter is generally labeled occupational health and well-being and has been related to organizational innovations, primarily through the concept of 'stress.'

Regarding the effect on well-being of exposure to technology, Salanova and Schaufeli (2000) argued that it affects users' mental health and well-being both positively and negatively. The introduction of new ICT systems can lead to increased well-being by empowering employees but, at the same time, it alters organizational structures and processes. It can have unintended negative impacts, such as increased dependence, decreased freedom and autonomy (Gripenberg, 2004; Tolsby, 2000), reduced social exchange, and isolation of individuals (Zuboff, 1988). Moreover, some authors have identified information overload as an important stressor (Burke, 2009), and it is difficult to think of anything generating more information than new communication technologies.

Starting in the late 1960s but mainly since the 1980s, many authors have studied the relationships between technological innovations and their negative impacts on attitudes, thoughts, behaviors, and psychology, embracing concepts such as technostress, technophobia, or computer stress (Wang, Shu, & Tu, 2008). For example, technostress is defined as 'one of the fallouts of an individual's attempts and struggles to deal with constantly evolving ICTs and the changing cognitive and social requirements' (Tarafdar et al., 2007, p. 303). Leaving aside terminological differences and taking a broader perspective, it seems clear such stress leads to less job satisfaction and greater anxiety, hence decreased well-being (Tarafdar et al., 2007).

Godard and Delaney (2000) argue there is undeniable evidence that new work systems and human resource management practices can cause stress. Whereas Ramsay and colleagues (2000) find evidence that they put employees under more strain, Landsbergis and colleagues (1999) link them to increased stress. With regards to flexible work arrangements, Russell and colleagues' (2009) study finds that those working from home experience more work pressure and Harris's (2003) investigation reports one-third of the employees in her case study felt home-based working had increased rather than reduced their stress levels.

Additionally, as noted by Green (2004a, p. 717) 'there is a growing body of case study evidence linking the modern transformation of workplaces with the experience of health hazards brought on by work intensification.' Following Schabracq and Cooper (1997), the consequences of work overload vary from stress to exhaustion, insomnia, fatigue, and, in the long run, can lead to illness.

Indeed, there is quite extensive literature on the relationship between work intensification and occupational health problems, especially when referring to work time (Gospel, 2003). Sparks and colleagues' (1997) meta-analytic study on extended work time literature indicates a positive and significant correlation between long hours and health problems (physical and psychological). In terms of intensification of effort, Truss (2001) found that when employees invest more effort in their tasks, stress and other health problems increase. Wichert (2002) indicates work pressure and effort are related to less sleep that, in turn, leads to psychosomatic problems and affects psychological well-being. Green (2004b) studies the evolution of work effort and well-being over the period of 1992 to 2001 in the U.K. and concludes there has been certain deterioration.

In sum, previous studies have found evidence to link HRM or ICT to decreased employee well-being (Salanova & Schaufeli, 2000; Salanova, Cifre, & Martin, 2004; Schabracq & Cooper, 1997; Sparks, Faragher, & Cooper, 2001). This has been hypothesized to stem from work overload, invasion of private life, high complexity, and constant renewal that produce tension, stress, and even burnout (Carayon, 2007; Tarafdar et al., 2007). As a result, we expect organizational innovations in our study (ICTs and flexible work arrangements) to have two types of effects on employees: a direct negative effect and an indirect one resulting from increased work intensity.

We must add that the literature also reports numerous positive effects of organizational innovations. However, we believe this is not necessarily incongruous or counter to our propositions (Robey & Boudreau, 1999). Whereas the introduction of ICTs in organizations may speed up communication and keep employees better informed, hence satisfied (Tolsby, 2000), it can also make them feel pressured and overwhelmed (Tarafdar et al., 2007). Similarly, individuals may not respond to the new HRM practices by simply being more motivated (Appelbaum et al., 2000) or making more effort (Green, 2004a), but by overexerting themselves, which could jeopardize their well-being. Indeed, Salanova and Schaufeli (2000, p. 385) emphasize that exposure to innovation 'influences users´ mental health and well-being in a positive as well as a negative sense.' Our study focuses on the latter because it has been overlooked so far, but the positive side should be kept in mind.

METHODOLOGY

With the objective of achieving a better understanding of the previously discussed complex links and of contributing to theory building, we use case study methodology (Boxall & Steeneveld, 1999; Gratton, Hope-Hailey, Stiles, & Truss, 1999; Hailey, Farndale, & Truss, 2005) and have designed our analysis based on a single in-depth case study, following the example of previous research (e.g., Donnelly, 2008; Thompson, 2007; Truss, 2001). The following paragraphs justify our case selection and describe this research setting and our research design.

Research Setting

We have chosen a multinational company with more than 200,000 employees that provides different kinds of business services (i.e., consulting, technology solutions, and outsourcing) to clients in different industries. Our choice of this company is justified by its wide experience in the targeted organizational innovations, which are cutting-edge flexibility enhancing HR practices and the continuous introduction of new ICTs. As a senior manager of the company highlighted:

> We get so used to the enormous amount of services that the company offers us and to the continuous introduction of innovations, in terms of

> new ICT tools, flexibility policies etc, that it seems that this is the normal situation in all companies. But it is not. (Female, Senior Manager)

Besides, most employees in the organization are consultants who are believed to represent the archetype of knowledge workers (Fincham, 1999). Because they are highly skilled and degree-educated, the firm is interested in both attracting and retaining them. However, this type of employee experiences greater work pressure when compared to other occupational levels (Gallie, 2005) and might, therefore, be expected to suffer from intense work and well-being problems. On the other hand, if in this case a process of work intensification and deterioration in well-being is not evidenced, then it can be regarded as extremely unlikely that such a process might occur significantly in other settings (Goldthorpe, 1968).

Moreover, the company is leader of its sector and provides services to over 1,000 clients (both large private businesses and the government sector at local, regional, and national levels) and has been awarded a number of prizes for its ability to innovate, generate, and diffuse new ideas and knowledge.

This chapter will report on the first findings of a long-term research project, which circumscribe to the company's brand in Spain, with more than 10,000 employees. In this sense, Wachter and colleagues (2006, p. 255) believe the Spanish subsidiaries of international firms have been 'a strategic test bed for the implementation of innovative HR practices' because of their 'malleability.' As noted by Rodriguez-Ruiz and Martinez-Lucio (2010, p. 135), international companies, such as this one, have been essential for implementing new HR practices and have been the origin of HR development in the country.

As a consequence, this is considered to be a critical case and believe that our results should provide valuable empirical and conceptual understanding, applicable in many other organizations (Goldthorpe, 1968).

Research Design

The selected firm is divided into four principal groups: three groups that are dedicated to providing the previously mentioned business services to clients ('front office') and a fourth that supplies internal services to the rest of the company ('back office'). We interviewed employees from the two areas, but our results concentrate on the last group, because it is only where flexible working practices have been officially implemented so far. Moreover, its responsibilities are somewhat standard for any organization, therefore its functioning and characteristics should be consistent across organizations.

The data collection for the case was based on qualitative methods (70 recorded interviews and document analysis). First, we analyzed the vast official written material (annual reports, HR policies, and internal HR communication documents), and had access to the company intranet to gather information on existing ICT tools and ICT policies. Second, we conducted three rounds of interviews:

1. First, we interviewed the head of HR and three members of her team, and then the head of the department managing the company's technology infrastructures. We discussed the extant HR policies and ICT strategy to have clearer idea of the official organizational discourse and to detect the tools and practices most interesting for our analysis.
2. Then, we held 26 narrative very open interviews with employees from all professional categories. By doing this, we overcame the problem of any gap between intended and implemented practices (Khilji & Wang, 2006), and the fact that manager's perceptions tend to be different to their subordinates' (Eaton, 2003). These interviews were very searching and helped us narrow down our variables by probing into the undesirable effects of organizational innovations.
3. Last, we focused exclusively on the 'back office' and conducted 39 semi-structured interviews with employees from different departments. All interviewees had been with the company for at least one year, so they had experienced work both before and after the introduction of the organizational innovations. At this point, our aim was to collect data on their experiences with the selected organizational innovations, both flexibility practices (i.e., flexitime and telework) and ICT implementation (i.e., the communication toolset, innovation tool, and travel tool), to contrast them with the discourse of the HR and technology departments' official policies. The profile of the interviewees (gender, marital status, parenthood, years in the organization, etc.) were varied to have as diverse a sample as possible.

In total, 70 employees were interviewed for this study. Nonetheless, as previously noted, the results presented in this chapter are limited to our findings within the 'back office' group; hence, our focus is on the third set of 39 interviews, although the earlier 31 interviews contributed to the development of the initial model of analysis.

FINDINGS

As already mentioned, we have explored the effects of the introduction of two types of organizational innovations: flexible working practices and new ICT tools. The former can be divided in two practices: (1) flexitime, a practice allowing daily changes in the working schedule, having a nine-hour minimum workday instead of fixed starting and finishing times; (2) telework, a practice giving employees the opportunity to work from home a maximum of two days per week. The latter, new ICT tools, is threefold: (1)

communication tool: a collaboration toolset that aims to enhance communication between physically separated employees, enabling them to share documents, make calls from computer to computer, have round-tables and meetings with various people, share the computer desk, and so on; (2) innovation tool: a tool aiming to trigger creativity and innovation within the company providing a channel for employees to share and discuss their innovative ideas (each year some winning ideas are selected for implementation); and (3) travel tool: an administrative self-enablement tool related to managing employee travel (e.g., flight reservations).

In this section, we present evidence of negative employee experiences with these organizational innovations, focusing on the two undesirable outcomes previously mentioned (work intensification and decreased employee well-being). More than half our interviewees indicated that flexible working or new ICT tools have somewhat increased work intensity, although only 20% characterized this as serious. Another 30% mentioned possible negative effects on their well-being (such as increased work pressure, stress, and anxiety), although most characterized these as 'mild,' and only 10% felt their well-being was significantly affected.

Work Intensification

The introduction of innovations is often related to creating new processes and methodologies or increasing the complexity of existing ones. Such new processes seem to imply a higher workload. Two alternative paths to work intensification emerge from the analysis of the interviews, each with a different effect on employee well-being.

Involuntary intensification. As previously discussed, implementing organizational innovations can change working patterns and routines, which result in workers exerting greater effort to complete their tasks. In line with the literature, our findings show three drivers of such an increase.

1) Distortion of work boundaries: First, the limits of working time and space are blurred. With traditional systems, work started at 9 a.m., finished at 7 p.m., and was confined to the worksite. Conversely, current work has become ubiquitous. Whereas most of our interviewees report to be very satisfied with flexitime, which allows them to balance work and personal responsibilities, they also point out that they work longer hours as a result of not having a pre-set work time.

> Flexibility means working longer hours because we don't count the time we spend at work. People who have a fixed schedule know how long they stay in the office. (Female, Senior Manager)

In parallel, the use of ICT enables work to happen everywhere: all employees in this organization have a laptop and a cell phone used for work in the office, from home, at airports, doctor's waiting rooms, their children's kindergarten, or the grocery store. Of our interviewees, 50% percent feel that because work time is flexible, they should be available and always contactable. Hence, they are never 100% off work.

This issue makes it hard for employees to disconnect while "off" work, which reduces the 'individual's sense of being away from the work situation' (Etzion, Eden, & Lapidot, 1998, p. 579).

> Because our emails synchronize with our cell phones we are hooked . . . before sometimes we took our computers home, but now we don't disconnect because we're continuously checking our emails on the phone. (Female, Senior Manager)

2) Increased work pace and multitasking: Second, we found some organizational innovations intensify the work pace and create new tasks to complete, thus requiring greater intensive and extensive effort. On the one hand, the communication tool (used by all our interviewees) has been described as 'invasive,' producing what Herbig and Kramer (1994) labeled as 'information overload.' This phenomenon is related to the 'real time' characteristic of the tool introduced, which has accelerated communication processes, making the 'right here right now' possible. It also means workers' responses to requests (e.g., information, assistance) have increased because of the improved communication system:

> You can decide when to answer the phone or an email, but with this tool the requests pop up constantly . . . you feel intimidated because it seems you need to answer immediately. (Female, Director)

On the other hand, the travel tool is included within a group of 'self-enablement' technologies that imply additional tasks to the employee and, hence, cause work intensification (e.g., before the introduction of this tool, plane tickets and hotel and car reservations were managed through a call center). Some interviewees highlighted that many of the new tools are geared toward this 'self-service,' justified by the current financial crisis and the need to cut costs, and not toward improving working processes. For example, travel tool assistance can be requested, but it is charged against the personal account (i.e., penalization). In other words, with this 'self-enablement' philosophy, employees now need to assume new tasks as their responsibility.

These problems increase when the innovations work inefficiently or their design is inadequate because time is wasted. As an interviewee expressed:

> Often, when you can do things for yourself efficiency improves. However, this will only happen if the tools function properly. If they don't, they create additional work. For example, the travel tool is badly designed and causes many problems. (Female, Manager)

3) Constant need for learning and adaptation: Our respondents constantly reiterated the need to train themselves and adjust to change. When a new organizational practice or tool is introduced, time must be spent learning how to use it, implying that this should be assumed as part of their work. With a new implementation, there is generally initial confusion, transition time, and a learning curve depending both on the complexity of the practice and the employee's personal characteristics, because 'not everybody assumes changes the same way,' the director of technology infrastructures says. Although around 35% of the interviewees considered it easy or very easy to adapt to the company's continuous innovation, and another 35% consider it quite easy, 30% still found it rather difficult. They commented on issues such as the lack of time to get updated due to the fast pace of innovation, the effort to train themselves and get used to innovations, and the increasing role of 'self-enablement' tools, such as the previously mentioned travel tool. On this point, one of these interviewees explained:

> The company counts on its employees' capabilities . . . we're used to these changes and adapt because we can. Nobody teaches us how to use the tools . . . the company assumes we'll manage without help. (Female, Manager)

Indeed, employees in this company are used to adjusting and adapting to innovations, and probably find it easier than most. Nevertheless, one interviewee pointed out that even in constantly innovating companies with a philosophy of continuous improvement, such as this one, 'there is a moment when employees reach the limit of their absorptive capacity' (Male, Senior Manager). All in all, the need to adapt engendered both greater extensive and intensive effort.

Voluntary intensification. Besides the previously described expected results, our data show that introducing certain organizational innovations also lead to workers' voluntarily exerting more effort. This is particularly true for users of the innovation tool and teleworkers.

Besides their regular tasks, the former decide to log in and contribute with their personal ideas and suggestions to the discussions in the forum. About 40% have used the tool to varying degrees. Because their regular workload is not reduced, extensive work effort increases as a result of the employees' willingness to participate.

> We're motivated to participate but we lack the time because of our demanding everyday work which is always dedicated to projects. Besides this, there's no recognition of the time invested in this tool. The time needed to think about improvements or innovative ideas must be taken from our own personal time. (Female, Manager)

As far as remote working is concerned, 80% of teleworkers reported greater intensive or extensive effort. They seem more prone to devoting longer hours, working both during the time they save from commuting and at nonregular working hours.

> I definitely work longer because I use the time that I'd be commuting for work. (Female, Specialist)

> The employees that work from home don't mind starting at 7 or even 6.30, or working late, for example having a conference call and accommodating international time differences. (Male, Manager)

Increased intensive effort was also reported as a result of teleworking. Interviewees felt they had fewer distractions at home, which allowed them to concentrate more and devote greater cognitive attention to their task.

> Concentration is better at home. People can't walk by to chat and I feel they are less inclined to call me at home than at the office. I also take shorter breaks. Having coffee is just about having coffee, not about socializing, so I'll only be away from my desk for 5 to 10 minutes, whereas in the office I may take up to half an hour. (Female, Analyst)

In addition, flexitime was also mentioned as a motivation for voluntary intensification. Because employees felt they could take time off if they needed to, they were inclined to work longer for the organization when the workload required them to do so.

> With flexitime I can arrive at the office at 9:30. This allows me to drive my daughter to school every day. It's something I value immensely and that makes me feel very grateful to the company. In return, it's true that 90% of days I stay longer than I should, but I don't mind. (Male, Manager)

This kind of work intensification is not negatively perceived by the workforce. It seems to be the result of increased commitment and a positive social exchange (Blau, 1964; Cropanzano & Mitchell, 2005). Indeed, one of the essential characteristics of employee commitment is a 'willingness to exert considerable effort on behalf of the organization' (Mowday, Steers, & Porter, 1979, p. 226). In the past, innovative HR practices have been proven to increase commitment (Gould-Williams & Davies, 2005; Whitener, 2001) because they are interpreted by workers as indicators of the organization's interest in them (Wayne, Shore, & Liden, 1997). In the same way, the innovation tool gives the employees an opportunity to participate in organizational progress and decision making, increasing the motivation of those who participated. Also, providing the employees with the option to work from home makes them grateful to the organization (Peters & van der Lippe, 2007) and encourages them to reciprocate in kind:

> Two days a week I can have all three meals with my family and avoid traffic because my organization trusts me and allows me to work from home. Of course I need to repay this with my effort. (Male, Manager)

Well-Being

Our findings support the proposition that innovative HRM and ICT can affect employee well-being. However, even if previous studies have linked work intensification to decreased well-being (Schabracq & Cooper, 1997; Wichert, 2002), our results are not entirely aligned with them. Indeed, we found that interviewees affected by voluntary work intensification were much less likely to point out any health and well-being problems than those affected by involuntary intensification.

This finding is theoretically supported by the job demands-job control model that focuses on the distinction of discretion (the extent to which employees have the potential to control what they do throughout the workday) and demands (stressors existing in the workplace) associated with jobs (Karasek, 1979). The central tenet of this theory is that employee discretion over work mitigates the negative effects job demands have. Hence, increasing demands without increasing discretion should have a significant negative impact on well-being.

In this way, the organizational innovations that generated involuntary intensification (i.e., that brought in additional demands) were identified as possible sources of distress and anxiety. For instance, when discussing flexitime, an interviewee said:

> Not having a real working schedule can become very stressful. At work I have to be 100% attentive, but when I leave the office I still have to be at least 50% in case something comes up. (Male, Specialist)

The communication tool was also singled out as a source of pressure and stress. As personal agendas are managed openly, anyone can see when another colleague has a free space and arrange new appointments for them. The tool shows who is at their desk and available at all times. Interviewees commented:

> You aren't the owner of your time anymore . . . productivity is much higher, but the rhythm can be stressful. (Male, Senior Manager)

In contrast, if organizational innovations empower employees and develop their sense of control, there seems to be no negative impact on well-being.

Again, in line with the job demands-job control model, when demands go along with increased discretion, no decrease in well-being is reported. For example, although, as previously noted, teleworkers report working longer more intensive hours, none of them felt working from home had a negative impact on their well-being.

The interviewees did not report the innovation tool as having an impact on work pressure either. However, we did find it had negative impacts on employee well-being resulting from three unintended effects. Some interviewees stated that, in some cases, (1) the introduction of this tool had created greater competitiveness and even jealousy among colleagues, because of the selection process of the best ideas and the resulting innovation 'winners.' Moreover, (2) some people got demotivated after proposing many ideas that were not selected. Besides, (3) the use of the tool is completely voluntary but most employees lack the time to use it, so they could think 'that people providing most of the ideas have more free time (time not dedicated to clients) and on top of that, they get awarded for their participation' said a female manager.

In addition, we must point out three contingency factors to the 'organizational innovations–well-being' relationship: (1) the adjustment time to the new innovation, (2) the employee's personality, and (3) the proportion of individual working versus interactive working time experienced. Regarding the first, the negative feelings (mainly stress) related to innovations is stronger initially, when they are introduced, and they then diminish as employees get used to them. Different interviewees explained:

> It requires discipline to learn how to use the new channels and to manage them . . . once you achieve this, the stress diminishes. (Male, Senior Manager)

> Having a flexible schedule may be hard to manage but one can learn. (Female, Senior Analyst)

The second factor, the employee's personality, includes a range of variables, of which two stand out. The previously mentioned worker's ability to disengage when outside working hours is the first. Whereas approximately 50% of the interviewees mentioned they never turned off their working phones, the other half willingly disengaged and reported well-being problems less often. The second relevant personal characteristic is the attitude toward organizational innovations, adaptation, and change. The happier people are to welcome change, the less likely it will adversely affect their well-being. Additionally, their previous experiences of dealing with change seem to have a positive influence.

The third contingency factor is the proportion of individual versus interaction time experienced. The fact that certain innovations, such as telework or the communication tool, diminished face-to-face interactions and interpersonal relations, reducing personal contact and friendly relationships, has made feelings of isolation arise. An interviewee said:

> Working from home is convenient, but I need to come to the office and have some kind of social interaction. Otherwise I just feel too alone. (Male, Specialist)

So it would seem that when a good equilibrium between individual work and social interaction is promoted, well-being is less likely to be affected by organizational innovations.

DISCUSSION

This study intended to identify some plausible negative effects of organizational innovations within the context of a pioneering company. In accordance with previous research, we have shown that these innovations can lead to both work intensification (Burchell et al., 2002; Green, 2004a) and problems of well-being.

First, we have documented the evidence of a double path to work intensification. On the one hand, some innovations affect working patterns and routines that require workers to make a greater all-round effort (Bonazzi, 1993). On the other hand, others support the workforce and boost their commitment (Eisenberger, Huntington, Hutchison, & Sowa, 1986), incentivizing employees to voluntarily increase their effort (Appelbaum et al., 2000).

Second, two intertwined factors have been found relevant to determining which pathway is triggered by the implementation of an organizational innovation: (1) perceived employee autonomy and sense of control and (2) perceived organizational support. These findings are summarized in figure 10.1.

1) Perceived employee autonomy and sense of control is linked to the voluntary adoption or use of the innovation. Teleworking is an example of this, being offered to employees but not imposed on them. Each individual

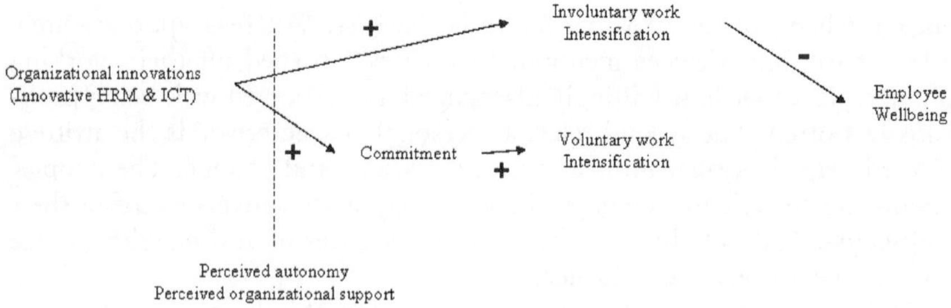

Figure 10.1 Schematic overview of relationships between organizational innovations and employee well-being.

can come to an agreement with their supervisor to work from home up to a maximum of two days. This process is initiated by the employees who therefore feel they have control over the innovation's implementation. The innovation tool is also used on a voluntary basis. Employees are encouraged although not obliged to use it, hence, they perceive their participation as an autonomous decision. In contrast, the use of the communication tool made some employees feel they are not the 'owners of their time anymore,' implying a diminished sense of control and the feeling they have to respond to petitions immediately, even if involuntarily. The case of flexitime is particularly enlightening on this point. The employees that showed greater commitment and reported to purposefully devote more effort to their job felt that flextime allowed them to better control their workday. In contrast, when flexitime workers felt less able to organize their day, involuntary intensification was described. In summary, it seems that feeling in control and empowered makes employees more likely to commit to the organization (Gellatly, Hunter, Currie, & Irving, 2009; Whitener, 2001), triggering the voluntary intensification pathway. Similarly, Lindell (chapter 8 in this volume) argues that the employees' involvement in implementing innovations may counteract their possible negative effects.

Additionally, we must point out that, as expected, the perception of autonomy seems linked to the employees' professional category: managers and senior managers feel much more in control of the implementation of organizational innovations than lower-level employees.

2) Perceived organizational support (POS), the second factor, is closely linked to the first. As noted by Eisenberger and colleagues (2004), organizational practices, such as the innovations this study evaluates, can noticeably contribute to POS because they are specifically oriented toward employees. Previous studies found that practices enhancing participation in decision making and autonomy were related to POS (Allen, Shore, & Griffeth, 2003; Rhoades & Eisenberger, 2002) if their implementation is perceived as intentional organizational action to benefit employees (Eisenberger, Jones, Aselage, & Sucharski, 2004, p. 211). The travel tool affects employees but is not intended to benefit them directly. In contrast, although teleworking may help the organization reduce office space cost, it is only perceived as an employee-oriented practice that is intended to balance the employees' work-life balance. Hence, it is more likely to generate voluntary work intensification as it contributes to POS, whereas the travel tool does not.

Third, in this study we identified that the link between work intensification and decreased well-being seems to depend on the pathway leading to such work intensification. In accordance with Karasek's (1979) postulates for the job demands-job control model, we found that when work intensifies involuntarily, feelings of anxiety and stress arise. In the case of voluntary intensification, the workers have discretion and a sense of control that mitigates the negative effect of increasing demands; hence, they do not feel their well-being is negatively affected. Nevertheless, even if intensification is voluntary, thus without the adverse effect on well-being in the short term, we agree with Green's (2001, p. 55) statement that 'a process of growth that relies on work intensification is sure to be self-limiting.' In this sense, identifying and controlling all kinds of work intensification may be advisable.

In addition, organizational innovations seem to have a direct impact on employee well-being that is contingent on three factors: the adjustment period to the innovation, the employee's personality, and the proportion of individual working time versus interactive working time experienced.

However, as previously noted, it is important to remember that, so far, we have only focused on reporting the negative or unintended effects of innovations, as was the objective of this chapter. This does not mean, however, that the negative effects outweigh the positive ones. In fact, the general opinion of interviewees is that they are positive: They value the technology and flexibility made available to them. Regarding the former, each of the ICT tools are considered to have specific advantages: (1) the communication tool has made communication easier, faster, cheaper, and so on, and has broken down geographical barriers; (2) the travel tool has cut administration costs; and (3) the innovation tool has fostered participation and led to the launching of some of the proposed ideas, increasing employee satisfaction. With regards to flexible working practices, employees consider balancing personal and work issues as the most beneficial outcome and also highlight that having people work from home reduces required office space. All in all, one cannot say that organizational innovations have negatively impacted on the company. Nevertheless, being aware of the previously described side-effects and managing them, for example, by providing training and support (Burke, 2009), should render their net effect even more beneficial.

Moreover, the negative effects on employee well-being do not seem to be an effect of innovation per se, but its management. For instance, the fact that people have flexibility regarding when they start and finish work should not be detrimental to employee well-being. However, using this flexibility to increase the workload without accounting for overtime is. Similarly, having communication tools is good idea, as it can improve information sharing and increase employee satisfaction. Nonetheless, having employees constantly monitoring and chasing each other is detrimental. Managing these issues is therefore crucial to the success of the innovations and generally implies first coaching, teaching, and encouraging individuals to ensure that they have the necessary skills and confidence (Mikkelsen, Øgaardb, Lindøea, & Olsen, 2002). Second, the implementation should be coherent with other company policies and the organization culture if the intended result in terms of its practice is to be achieved (Orlikowski, 2000). To avoid negative side effects it is not sufficient to provide employees with numerous tools for various purposes, it is also necessary to give them the time to use them and avoid wasted efforts caused by innovation overload.

According to at least 85% of interviewees, the company seems conscious of the potential negative impact these innovations could have at all levels and evaluates their pros and the cons. However, some employees have stated that this awareness is limited to the direct short-term effects, and secondary or long-term effects are not taken into account. For example, the company is aware that the innovation tool has a positive short-term or direct impact on motivation because employees feel that their ideas are appreciated. However, we must add that, over time, the innovation tool appears to be causing a secondary or long-term negative impact because it can give rise to envy or greater competitiveness. This secondary impact is not contemplated by the company. Nevertheless, being aware of an innovation's indirect and long-term effects on employees seems to be essential and constitutes the first step to managing the problem.

LIMITATIONS AND NEXT STEPS

First, we must note that the analysis conducted in this chapter is only exploratory. However, the quantitative testing of the presented model, using an employee random survey, will be the next step in our research project.

Second, although the analysis is limited to one company, authors such as Donnelly (2008), Thompson (2007), Truss (2001), and Hailey and colleagues (2005) have made significant contributions in the innovative HRM domain by thoroughly analyzing one specific firm and understanding its processes. In addition, the literature has called for further qualitative in-depth research in this field (Guest, 1997; Purcell, 1999). Nonetheless, the results will be revisited by replicating the analysis in other companies. Moreover, even if the results cannot be directly extrapolated to other organizations, as noted in the methodology section, this is a valuable study because it is looking at a critical case (Goldthorpe, 1968).

Third, it is well-known that the innovation process is industry-biased (Malerba, 2004; Castellacci, 2008), and as our analysis focuses on just one industry, results would have to be contrasted with other industries.

Fourth, many innovative HRM practices or work organization ICTs have not been considered (e.g., task flexibility, new compensation systems, tools for work-time reports, virtual communities, customer-management relationship software, etc.), as it was essential to focus on a selected number of innovations to allow us to draw useful conclusions for management purposes.

Finally, the elaboration of policy recommendations for companies is beyond the scope of this chapter. A next step in our research will be to explore the company's awareness of the contemplated undesirable consequences and identifying to what extent they are being addressed. This will be a first step in establishing guidelines to maximize the relationship between organizational innovations and performance.

CONCLUSION

Debates over the extent to which organizational innovations have negative consequences on employees have been extremely rare. This chapter attempts to fill this gap in the literature by examining the detrimental effects of two types of organizational innovations: flexibility enhancing innovative HRM practices and new ICT tools. By means of an exploratory case study, we have found that these innovations can intensify work and reduce employee well-being.

Our results show that organizational innovations tend to increase work time and work effort, in other words, they generate work intensification. In accordance with the literature, we find three drivers that explain this increase. Organizational innovations: (1) distort work boundaries, making work happen all day long, everywhere; (2) induce multitasking and accelerate the work pace; and (3) require employees to constantly adapt, change, and learn. These issues oblige employees to work longer and more intensely; hence we label this phenomenon involuntary intensification.

Besides these reasons for work intensification, we find it is also experienced when organizational innovations to empower or support employees are introduced. However, in this case, workers feel in control and the increased work time and effort are a consequence of their desire to contribute to the firm in return for the interest shown in them: work intensification happens voluntarily, resulting from social exchange.

This study has also found that depending on the pathway leading to work intensification, increased work time and effort may or may not impact on employee well-being. Whereas involuntary work intensification leads to greater stress, work pressure, and feelings of anxiety, voluntary work intensification does not seem to negatively impact well-being.

Beyond work intensification, we find three other variables that affect the link between organizational innovations and decreased well-being: (1) The adjustment period to the innovation: the more time, the weaker the link; (2) the employees' personality: the greater their ability to disengage from work, the more positive their attitude toward organizational innovations. Also, the greater their experience at dealing with change, the less strong the link is between organizational innovation and well-being; and (3) the proportion of time spent working alone compared to time spent interacting with colleagues: When time spent alone exceeds a certain level, employees are more likely to experience problems of well-being.

All in all, it is important for managers and HR practitioners to consider the impact that different organizational innovations have on employees. If the nature and implementation of the innovations are likely to negatively influence employee well-being, it is reasonable to assume that they will also reduce their performance. Although organizational innovations do have very positive impacts, overlooking these issues may be limiting their potential. Tackling them in time may counteract the negative consequences and allow the building of sustainable positive processes.

ACKNOWLEDGEMENTS

Almudena Cañibano would like to acknowledge the funding support received from the 'Fundación Banco Herrero' for this research. The authors would like to thank Mr. JPM, member of the company's executive board, for kindly granting us access to the organization, as well as all the employees that have shared their experiences with us.

*Several elements of this chapter are part of the doctoral theses of Almudena Cañibano and Oihana Basilio.

REFERENCES

Akerlind, I., & Schunder, S. (2007). Perspectives on health. In U. Johanson, G. Ahonen, & R. Roslender (Eds.), *Work health and management control.* Stockholm: Thomson Fakta.

Allen, D. G., Shore, L. M., & Griffeth, R. W. (2003). The role of perceived organizational support and supportive human resource practices in the turnover process. *Journal of Management*, 29(1), 99.

Appelbaum, E., Bailey, T., & Kalleberg, A. L. (2000). *Manufacturing advantage: Why high-performance work systems pay off.* Ithaca, NY: Cornell University Press.

Ash, J. S., Berg, M., & Coiera, E. (2004). Some unintended consequences of information technology in health care: The nature of patient care information system-related errors. *Journal of the American Medical Informatics Association,* 11(2), 104–112.

Avgerou, C., Ciborra, C., & Land, F. (Eds.) (2004). *The social study of information and communication technology: Innovation, actors, and contexts.* Oxford: Oxford University Press.

Bayo-Moriones, A., Billón, M., & Lera-López, F. (2008). Skills, technology and organisational innovation in Spanish firms. *International Journal of Manpower,* 29(2), 122–145.

Bircher, J. (2005). Towards a dynamic definition of health and disease. *Medicine, Health Care and Philosophy*, 8, 335–341.

Black, S., & Lynch, L. (2001). How to compete: The impact of workplace practices and information technology on productivity. *Review of Economics and Statistics,* 83, 434–445.

Blau, P. M. (1964). *Exchange and power in social life.* New York: John Willey.

Bonazzi, G. (1993). Modelo japonés, toyotismo, producción ligera: Algunas cuestiones abiertas. *Sociología del Trabajo*, 18, 3–22

Boxall, P., & Steeneveld, M. (1999). Human resource strategy and competitive advantage: A longitudinal study of engineering consultancies. *Journal of Management Studies*, 36(4), 443–463.

Brynjolfsson, E. (1993). The productivity paradox of information technology. *Communications of the ACM*, 36(12), 66–77.

Brynjolfsson, E., & Hitt, L. (1995). Information technology as a factor of production: The role of differences among firms. *Economics of Innovation and New Technology*, 3(4), 183–200.

Burchell, B. J., Lapido, D., & Wilkinson, F. (Eds.) (2002). *Job insecurity and work intensification.* New York: Routledge.

Burke, M. S. (2009). The incidence of technological stress among baccalaureate nurse educators using technology during course preparation and delivery. *Nurse Education Today*, 29(1), 57–64.

Carayon, P. (2007). Healthy and efficient work with computers and information and communications technology—Are there limits? *Scandinavian Journal of Work Environment & Health*, 3, 10–16.

Carter, R., Danford, A., Howcroft, D., Richardson, H, Smith, A., & Taylor, P. (2010, June). *Occupational ill-health and absence management under a lean regime in the UK civil service*. Paper presented at the Work Employment and Society Conference, Brighton.

Castellacci, F. (2008). Technological paradigms, regimes and trajectories: Manufacturing and service industries in a new taxonomy of sectoral patterns of innovation. *Research Policy*, 37, 978–994.

Ciborra, C. (2004). Encountering information systems as a phenomenon. In C. Avgerou, C. Ciborra, & F. Land (Eds.), *The social study of information and communication technology: Innovation, actors, and contexts*. Oxford: Oxford University Press.

Cropanzano, R., & Mitchell, M. S. (2005). Social exchange theory: An interdisciplinary review. *Journal of Management*, 31(6), 874–900.

Daniels, K., Lamond, D., & Standen, P. (2001). Teleworking: Frameworks for organizational research. *Journal of Management Studies*, 38(8), 1151–1185.

De Saá-Pérez, P., & Díaz-Díaz, N. (2010). Human resource management and innovation in the Canary Islands: An ultra-peripheral region of the European Union. *International Journal of Human Resource Management*, 21(10), 1649–1666.

Donnelly, R. (2008). Careers and temporal flexibility in the new economy: An Anglo-Dutch comparison of the organisation of consultancy work. *Human Resource Management Journal*, 18(3), 197–215.

Eaton, S. C. (2003). If you can use them: Flexibility policies, organizational commitment, and perceived performance. *Industrial Relations*, 42(2), 145.

Eisenberger, R., Huntington, R., Hutchison, S., & Sowa, D. (1986). Perceived organizational support. *Journal of Applied Psychology*, 71(3), 500–507.

Eisenberger, R., Jones, J., Aselage, J., & Sucharski, I. (2004). Perceived organizational support. In J. A.-M. Coyle-Shapiro, L. M. Shore, M. S. Taylor, & L. E. Tetrick (Eds.), *The employment relationship: Examining psychological and contextual perspectives* (pp. 206–225). Oxford: Oxford University Press.

Etzion, D., Eden, D., & Lapidot, Y. (1998). Relief from job stressors and burnout: Reserve service as a respite. *Journal of Applied Psychology*, 83(4), 577–585.

Fincham, R. (1999). The consultant-client relationship: Critical perspectives on the management of organizational chance. *Journal of Management Studies*, 36(3), 335–351.

Fleetwood, S. (2007). Why work-life balance now? *International Journal of Human Resource Management*, 18(3), 387–400.

Fleetwood, S., & Hesketh, A. (2006). HRM-performance research: Under-theorized and lacking explanatory power. *International Journal of Human Resource Management*, 17(12), 1977–1993.

Gallie, D. (2005). Work pressure in Europe: 1996–2001: Trends and determinants. *British Journal of Industrial Relations*, 43(3), 351–375.

Gardner, S. D., Lepak, D. P., & Bartol, K. M. (2003). Virtual HR: The impact of information technology on the human resource professional. *Journal of Vocational Behavior*, 63(2), 159–179.

Gellatly, I. R., Hunter, K. H., Currie, L. G., & Irving, P. (2009). HRM practices and organizational commitment profiles. *International Journal of Human Resource Management*, 20(4), 869–884.

Gerring, J. (2007). *Case study research: Principles and practice.* Cambridge: Cambridge University Press
Godard, J., & Delaney, J. T. (2000). Reflections on the 'high performance' paradigms implications for industrial relations as a field. *Industrial and Labor Relations Review*, 53(3), 482–502.
Goldthorpe, J. (1968). *The affluent worker.* Cambridge: Cambridge University Press.
Gospel, H. (2003). *Quality of working life: A review on changes in work organization, conditions of employment and work like arrangements.* Conditions of Work and Employment Series No. 1, International Labour Organization, Geneva.
Gould-Williams, J., & Davies, F. (2005). Using social exchange theory to predict the effects of HRM practice on employee outcomes. *Public Management Review*, 7(1), 1–24.
Gratton, L., Hope-Hailey, V., Stiles, P., & Truss, C. (1999). Linking individual performance to business strategy: The people process model. *Human Resource Management*, 38(1), 17.
Green, F. (2001). It's been a hard day's night: The concentration and intensification of work in late twentieth-century Britain. *British Journal of Industrial Relations*, 39, 53–80.
Green, F. (2004a). Why has effort become more intense? *Industrial Relations,* 43(4), 709–741.
Green, F. (2004b). Work intensification, discretion, and the decline in well being at work. *Eastern Economic Journal*, 30(4), 615–625.
Green, F. (2006). *Demanding work: the paradox of job quality in the affluent economy.* Princeton, NJ: Princeton University Press.
Green, F., & McIntosh, S. (2001). The intensification of work in Europe. *Labour Economics,* 8, 291–308.
Green, K. W., Wu, C., Whitten, D., & Medlin, B. (2006). The impact of strategic human resource management on firm performance and HR professionals' work attitude and work performance. *International Journal of Human Resource Management*, 17(4), 559–579.
Gripenberg, P. (2004). Virtualizing the office: Micro-level impacts and driving forces of increased ICT use. In T. Heiskanen & J. Hearn (Eds.), *Information society and the workplace: Spaces boundaries and agency.* London: Routledge.
Gripenberg, P. (2005). *ICT and the shaping of society: Exploring human-ICT relationships in everyday life* (Publication No. 143). Doctoral thesis, Swedish School of Economics and Business Administration, Helsinki
Guest, D. (1997). Human resource management and performance: A review and research agenda. *The International Journal of Human Resource Management,* 8(3), 263–276.
Hailey, V., Farndale, E., & Truss, C. (2005). The HR department's role in organisational performance. *Human Resource Management Journal*, 15(3), 49–66.
Harris, L. (2003). Home-based teleworking and the employment relationship. *Personnel Review*, 32(4), 422–438.
Herbig, P. A., & Kramer, H. (1994). The effect of information overload on the innovation choice process. *Journal of Consumer Marketing,* 11(2), 45–64.
Higgins, C., Duxbury, L., & Johnson, K. L. (2000). Part-time work for women: Does it really help balance work and family? *Human Resource Management,* 39(1), 17–33.
Hill, J. E., Grzywacz, J. G., Allen, S., Blanchard, V. L., Matz-Costa, C., Shulkin, S., et al. (2008). Defining and conceptualizing workplace flexibility. *Community, Work & Family*, 11(2), 149–163.

Hollenstein, H. (2004). Determinants of the adoption of information and communication technologies: An empirical analysis based on firm-level data for the Swiss business sector. *Structural Change and Economic Dynamics,* 15, 315–342.

Huselid, M. (1995). The impact of human resource management practices on turnover, productivity, and corporate financial performance. *Academy of Management Journal,* 38(3), 635–672.

Hyman, J., & Summers, J. (2004). Lacking balance? Work-life employment practices in the modern economy. *Personnel Review,* 33(4), 418–429.

Karasek, R. A., Jr. (1979). Job demands, job decision latitude, and mental strain: Implications for job redesign. *Administrative Science Quarterly,* 24(2), 285.

Khilji, S. E., & Wang, X. (2006). 'Intended' and 'implemented' HRM: The missing linchpin in strategic human resource management research. *The International Journal of Human Resource Management,* 17(7), 1171–1189.

Landsbergis, P., Cahill, J., & Schanll, P. (1999). The impact of lean production and related new systems of work organization on worker health. *Journal of Occupational Health Psychology,* 4, 108–130.

Lepak, D., & Shaw, J. (2008). Strategic HRM in North America: Looking to the future. *International Journal of Human Resource Management,* 19(8), 1486–1499.

Liker, J. K., Haddad, C. J., & Karlin, J. (1999). Perspectives on technology and work organization. *Annual Review of Sociology,* 25, 575–596.

Malerba, F. (2004). Sectoral systems: How and why innovation differs across sectors. In J. Fagerberg, D. Mowery, & R. Nelson, (Eds.), *Oxford handbook of innovation* (pp. 380–406). Oxford: Oxford University Press.

Marsden, D., & Cañibano, A. (2010). An economic perspective on employee participation. In A. Wilkinson, P. J. Gollan, M. Marchington, & D. Lewin, (Eds.), *The Oxford handbook of participation in organizations* (pp. 131–166). Oxford: Oxford University Press.

Mikkelsen, A., Øgaardb, T., Lindøea, P. H., & Olsen, O. E. (2002). Job characteristics and computer anxiety in the production industry. *Computers in Human Behavior,* 18(3), 223–239.

Mowday, R., Steers, R., & Porter, L. (1979). The measurement of organizational commitment. *Journal of Vocational Behavior,* 14(2), 224–247.

Oliner, S. D., & Sichel, D. E. (1994). Computers and output growth revisited: How big is the puzzle? *Brookings Papers on Economic Activity,* 2, 273–334.

Organisation for Economic Co-operation and Development. (2005). *The Oslo manual.* Paris: Organisation for Economic Co-operation and Development.

Orlikowski, W. J. (2000). Using technology and constituting structures: A practice lens for studying technology in organizations. *Organization Science,* 11(4), 404–428.

Peters, P., & van der Lippe, T. (2007). The time-pressure reducing potential of telehomeworking: The Dutch case. *International Journal of Human Resource Management,* 18(3), 430–447.

Pfeffer, J. (2010). Building sustainable organizations: The human factor. *Academy of Management Perspectives,* 24(1), 34–45.

Purcell, J. (1999). Best practice and best fit: Chimera or cul-de-sac? *Human Resource Management Journal,* 9(3), 26–41.

Ramsay, H., Scholarios, D., & Harley, B. (2000). Employees and high performance work systems: Testing inside the black box. *Journal of Industrial Relations,* 38, 501–531.

Rhoades, L., & Eisenberger, R. (2002). Perceived organizational support: A review of the literature. *Journal of Applied Psychology,* 87(4), 698–714.

Robey, D., & Boudreau, M. C. (1999). Accounting for the contradictory organizational consequences of Information Technology: Theoretical directions and methodological implications. *Information Systems Research*, 10(2), 167–185.

Rodriguez-Ruiz, O., & Martinez-Lucio, M. (2010). The study of HRM in Spain: The Americanization of Spanish research and the politics of denial? *International Journal of Human Resource Management*, 21(1), 125–143.

Russell, H., O'Connell, P. J., & McGinnity, F. (2009). The impact of flexible working arrangements on work–life conflict and work pressure in Ireland. *Gender, Work & Organization*, 16(1), 73–97.

Salanova, M., Cifre, E., & Martin, P. (2004). Information technology implementation styles and their relation with workers' subjective well-being. *International Journal of Operations & Production Management,* 24(1), 42–54.

Salanova, M., & Schaufeli, W. (2000). Exposure to information technology and its relation to burnout. *Behaviour and Information Technology,* 19(5), 385–392.

Schabracq, M., & Cooper, C. (1997). Flexibility of labour, well being and stress. *International Journal of Stress Management*, 4(4), 259–274.

Sparks, K., Cooper, C., Fried, Y., & Shirom, A. (1997). The effect of hours of work on health: A meta-analytic review. *Journal of Occupational and Organizational Psychology*, 70, 391–408.

Sparks, K., Faragher, B., & Cooper, C. L. (2001). Well-being and occupational health in the 21st century workplace. *Journal of Occupational & Organizational Psychology*, 74(4), 489.

Tan Cheng, L., & Nasurdin, A. D. (2010). Human resource management practices and organizational innovation: An empirical study in Malaysia. *Journal of Applied Business Research*, 26(4), 105–116.

Tarafdar, M., Tu, Q., Ragu-Nathan, B. S., & Ragu-Nathan, T. S. (2007). The impact of technostress on role stress and productivity. *Journal of Management Information Systems,* 24(1), 301–328.

Thompson, M. (2007). Innovation in work practices: A practice perspective. *International Journal of Human Resource Management*, 18(7), 1298–1317.

Tolsby, J. (2000). Taylorism given a helping hand: How IT systems changed employees' flexibility and personal involvement in their work. *Journal of Organizational Change Management*, 13(5), 482–492.

Truss, C. (2001). Complexities and controversies in linking HRM with organizational outcomes. *Journal of Management Studies*, 38(8), 1121–1149.

Vaast, E., & Walsham, G. (2005). Representations and actions: The transformation of work practices with IT use. *Information and Organization*, 15, 65–89.

Wachter, H., Peters, R., Ferner, A., Gunnigle, P., & Quintanilla, J. (2006). The Role of international personnel function. In P. Almond, & A. Ferner (Eds.), *American multinationals in Europe: Managing employment relations across national borders* (pp. 248–270). Oxford: Oxford University Press.

Wang, K., Shu, Q., & Tu, Q. (2008). Technostress under different organizational environments: An empirical investigation. *Computers in Human Behavior,* 24, 3002–3013.

Wayne, S. J., Shore, L. M., & Liden, R. C. (1997). Perceived organizational support and leader-member exchange: A social exchange perspective. *Academy of Management Journal*, 40(1), 82–111.

White, M., Hill, S., McGovern, P., Mills, C., & Smeaton, D. (2003). 'High-performance' management practices, working hours and work-life balance. *British Journal of Industrial Relations*, 41(2), 175–195.

Whitener, E. (2001). Do 'high commitment' human resource practices affect employee commitment? A cross-level analysis using hierarchical linear modelling. *Journal of Management*, 27, 515–535.

Wichert, I. (2002). Job insecurity and work intensification: The effects on health and well-being. In B. J. Burchell, D. Lapido, & F. Wilkinson (Eds.), *Job insecurity and work intensification*. New York: Routledge.

World Health Organization. (1948). Preamble to the Constitution of the World Health Organization as adopted by the International Health Conference, New York, 19–22 June, 1946; signed on 22 July 1946 by the representatives of 61 States (Official Records of the World Health Organization, no. 2, p. 100) and entered into force on 7 April 1948. Geneva: Author.

Zuboff, S. (1988). *In the age of the smart machine: The future of work and power*. New York: Basic Books.

11 Information and Communication Technology as an Exporter of CO_2 Emissions

Mitsutaka Matsumoto and Kotaro Kawajiri

This chapter focuses on the impact of information and communication technology (ICT) innovation on carbon dioxide (CO_2) emissions. The problem of CO_2 emissions or that of climate change exemplifies the unintended consequences of innovation because it is caused by the affluence of our lives achieved through innovation. The importance of the CO_2 emission problem has been recognized in recent years, and as a result, CO_2 impact assessment is sometimes attempted for new products and innovation. However, the impact of innovation is often extensive, causing indirect and unintended effects. In this chapter, we consider the impact of ICT innovation on CO_2 emissions as an example for discussing the indirect and unintended effects of innovation. In existing studies of ICT's CO_2 impact, the "direct" effects of ICT on CO_2 emissions have been addressed, assessing either the amount of electricity consumption increase caused by the expansion of ICT equipment use or energy saving resulting from the expansion of ICT applications. This chapter highlights the indirect effects of ICT on CO_2 emissions, an original analysis. The authors focus on the socioeconomic structural change caused at least partially by ICT and assess the resulting impact of CO_2 emissions caused by this socioeconomic change. To understand the net impact of innovation, it is necessary to assess not only direct but also indirect effects. In this chapter, first, several previous studies on ICT's impact on CO_2 emissions (the "direct" impact) are reviewed. Then, we discuss ICT's impact on socioeconomics and examine the impact of resulting socioeconomic changes on CO_2 emissions (the "indirect" impact) through several types of statistical data and data analyses. Finally, we discuss the implications of these analyses for understanding innovation's impact on CO_2 emissions.

ICT'S DIRECT IMPACT ON CO_2 EMISSIONS

ICT is one of the most influential innovations in the last few decades. The assessments of ICT's impact on CO_2 began after growing concerns regarding the increased use of ICT equipment causing an increase in electricity demand. A review of previous studies on ICT's direct impact on CO_2 follows.

ICT has a large impact on CO_2 emissions due to its high and increasing electricity demand. Estimates of ICT products' electricity consumption indicate that the electricity consumption of ICT equipment rages from 3% to 6% of a country's total electricity consumption, and that the share is increasing rapidly. For example, the ICT equipment in the U.S. was estimated to consume 74 billion kilowatt hours (kWh) in 1999 (Kawamoto et al., 2002) or 98 billion kWh in 2000 (Roth, 2005). ICT caused 2.0% of total CO_2 emissions in Japan in 2005 (Ministry of Internal Affairs and Communications, 2008) and around 2% worldwide, an impact similar to the aviation industry (Gartner Research, 2008).

The positive effects of ICT, or ICT's role in reducing CO_2, have also been studied along with the negative effects of ICT on CO_2. The results of the studies classify ICT's role in reducing CO_2 into four categories.

1. Improvement in energy efficiency
2. De-materialization
3. Substitution of transportation
4. Enhancement of environmental consciousness

The first factor indicates that ICT systems could support production management, supply chain management, and energy use management, all of which could improve the CO_2 efficiencies of production and consumption, thus reducing overall CO_2 emissions. Origuchi and colleagues (2007) stated that supply chain management supported by ICT enhances energy efficiency. The second factor is ICT's "de-materialization" effects. ICT could convert information media, such as books and documents, from paper to electronic media. Digitizing paper-based documents promotes "paperless" usage and could thus reduce CO_2 emissions in the paper-manufacturing industry. An encyclopedia, for example, which once consisted of approximately 20 paper-based volumes, today usually consists only of several CD-ROMs (Romm, Herrmann, & Rosenfeld, 1999) and can even be accessed via the Internet with no intervening media needed.

The third factor is the substitution of telecommunications for transportation. Web-based conferences and online shopping could have the effect of substituting telecommunication for transportation, reducing CO_2 emissions (Caudill, Luo, Wirojanagud, & Zhou, 2000; Williams & Tagami, 2002; Nakamura, Takahashi, Nishi, Kamekura, & Takeshima, 2000; Toffel & Harvath, 2004; Ueda, Nakazawa, & Hashitani, 2004). The fourth factor is that ICT could enhance the environmental consciousness of people by providing environmental load-related information. For example, car drivers provided with fuel efficiency information on their driving in real time become more conscious of the fuel efficiency of their driving and tend to drive in a more fuel-efficient manner. In an experimental project in Japan, an average improvement of 6% in fuel efficiency was observed.

Table 11.1 Previous Studies on ICT's Direct Impact on CO_2

Institutes/Authors	Area/Year	CO_2 ratio/ increase	CO_2 decrease	CO_2 Total
Denki Tsusin Shingikai (1998)	Japan 2010	—	-1.7%	—
NTT (2001)	"	+1.1%[5]	-3.6%	-2.5%
NEC (2003)[1]	"	+1.5%[6]	-4.6%	-3.1%
Shonan Econometrics (2004)[2]	"	—	—	-1.9%
Ministry of Internal Affairs and Communications (2005)	"	+0.4%[7]	-2.4%	-2.0%
Ministry of Internal Affairs and Communications (2008)	"	+2.3%[8]	-4.8%	-2.5%
CEPP[3] (2002)	Japan 2020	—	—	-0.08%[9]
Rodriguez et al. (2004)[4]	E.U. 2020	—	—	+2%~ -15%[9]

[1]Source: Matsumoto, Hamano, Tamura, & Iguchi, 2003.
[2]Source: Takase & Murota, 2004.
[3]CEPP: Energy sougo suisin inkai.
[4]Source: Rodriguez, Wunnik, Sancho, Burgelman, & Desruell, 2004.
[5]Ratio of CO_2 emission by ICT to the total CO_2 emissions in Japan in 1999.
[6]Ratio of the increase of CO_2 emission by ICT during 1995 and 2010 to the total CO_2 emissions in Japan in 1995.
[7]Ratio of the increase of CO_2 emission by ICT during 1990 and 2010 to the total CO_2 emissions in Japan in 2000.
[8]Ratio of CO_2 emission by ICT in 2010 to the total CO_2 emissions in Japan in 1990.
[9]The estimate of CEPP was represented by the amount of energy demand (unit: kcal) rather than the amount of CO_2 emissions. Similarly, the estimate in Rodriguez et al. (2004) was represented by the amount of green house gas emissions.

In addition to these four factors, a "smart grid," which functionally requires ICT, may also have a large impact on CO_2 reduction in the future (Friedman, 2008). A smart grid matches the demand and supply of electricity. It can smooth time fluctuations in electricity demand and suppress the peaks of demand, reducing CO_2 emissions in electric power plants. ICT's potential to reduce CO_2 at the national level has been studied since the late 1990s. Table 11.1 lists several relevant studies. The studies are based on a life cycle assessment (LCA) of ICT systems and admit that there is considerable uncertainty about the amount of CO_2 emissions reduced. According to the report of a committee of Ministry of Internal Affairs and Communications in 2008, for example, it was estimated that although ICT will cause 2.3% of CO_2 emissions in Japan in 2010, it will have the effect of reducing CO_2 by 4.8%, with a net effect of a 2.5% reduction in CO_2 emissions. Therefore, the spread of ICT equipment and applications is estimated to have both positive and negative effects on CO_2 emissions, with a net effect of decreased CO_2 emissions.

ICT'S SOCIOECONOMIC IMPACTS

Although the increase in ICT equipment's electricity demand is of concern and ICT application expansion's positive effects on CO_2 emissions are expected, ICT's socioeconomic effects far exceed the mere penetration of ICT equipment into all aspects of society and business. Analogous to the 18th century Industrial Revolution, which changed the structure of production, the form of labor, the location of industries, and the distribution of wealth and power in society, ICT has caused structural changes in the global economy. ICT's socioeconomic effects possibly cause a large amount of "indirect" effects on CO_2 emissions in society. Thus, to discuss the net impact of ICT on CO_2 emissions, we need to examine the nature of ICT's socioeconomic effects.

To understand the nature of ICT's structural effects, it is useful to identify newly emerging business forms. One good example of a new business form is offshore outsourcing, the migration of jobs from developed countries to newly industrialized countries. Countries trade to exploit their comparative advantages. Patterns of comparative advantage can change over time, and it is critical to understand offshoring in the ICT impact context. At one time, the U.K. had a comparative advantage in textile manufacturing; this advantage later shifted to the U.S., and jobs moved from the U.K. to U.S. Then, the comparative advantage shifted to Japan, and the advantage in textile manufacturing currently resides in China and other low-wage countries.

Recently, international division of labor has increased, and trade among countries has increased even more. This increase resulted in part from international political factors, whereby a new labor force comprising billions of people joined the market economy after the end of the Cold War, and in part from technological factors such as ICT, transportation systems, and product design rules (Baldwin & Clark, 2000). ICT has played a significant role in promoting the trend of economic globalization. ICT enabled the worldwide communication and reduced the barriers and costs for companies to operate offshore. Since the early 1990s, companies in developed countries have looked offshore to contain their labor costs, secure new sources of talent, and more quickly bring their products to market (Aird & Sappenfield, 2009). In 1990s, software companies in the U.S., for example, began adopting offshoring practices, using teams in different countries to quickly and efficiently develop code. This was enabled by robust ICT infrastructures that allowed for fast data transfer. The finance and insurance industries also began adopting offshoring strategies, as companies took advantages of the emerging technology for efficient voice-call routing through the Internet. More companies from different industries began making global sourcing part of their business strategies.

In this process, ICT has expanded the scope of goods and services traded across the borders. Economic theorists categorize goods and services as tradable or nontradable. Traditionally, any item that could be put in a box and shipped (manufactured goods) was considered tradable, and anything that could not be put in a box (services) or was too heavy to be shipped was considered nontradable. Improvements in technology have changed the boundary between what is tradable and what is not, with more items becoming tradable (Blinder, 2006). The old assumption that items that cannot be put in a box are nontradable is now obsolete. Because packets of digitized information act as virtual boxes, many services are now tradable. One example of such services is call center outsourcing. When an American consumer calls a company about a product, this call may connect to an agent in India. The consumer is not aware of this routing of the call. Some Japanese businesses use call centers in China. Offshore outsourcing became possible because ICT advances significantly lowered communication costs with these low-wage countries. Table 11.2 shows examples of offshore outsourcing.

Blinder (2006) calls exportable or importable services as "impersonal services" and those that cannot be provided without coming into contact with the client and thus cannot be exported or imported, including services such as those provided by caregivers, policemen, and surgeons, as "personal services." According to Blinder, approximately one-third of the American workforce of about 150 million people engaged in manufacturing and impersonal service will be subject to international competition in the future, and these jobs may be lost to developing countries. Trade among nations will increase further in the global economy as international division of labor intensifies. Not only in the U.S. but also in other developed countries, much of the workforce will be subject to international competition. In addition, industries, employment, and CO_2 emissions are transferred to the developing countries in recent times. In the next section, the effects of CO_2 exports on total CO_2 emissions in the world are explored.

Table 11.2 Examples of Offshore Outsourcing

Task	Countries (from)	Countries (to)
Call center service	U.S. E.U. Japan	India Iceland, Eastern Europe China
Software development	U.S.	India
Data input	U.S.	India, Philippines
CT scan data analyses	U.S.	India, Australia
Tutor	U.S.	India
Accounting	U.S. Japan	India India, China

Source: Friedman, 2005.

ICT'S INDIRECT IMPACT ON CO_2 EMISSIONS

Let us denote ICT's effect of migrating CO_2 emissions from developed countries to developing countries as a result of industry migrations as "*CO_2 emission exports,*" which we regard as an indirect effect of ICT on CO_2 emissions. In the following sections, the impact of CO_2 emission exports on the total CO_2 emissions in the world is assessed by reviewing relevant data.

Related Studies

There are few discussions of the effect of international division of labor on CO_2 emissions. Regarding its effect on society and the economy, some studies provided positive arguments that globalization (international division of labor) will promote global economic development (e.g., Smith, 2000), but others emphasized its negative aspects, such as the destruction of tradition and the widening of the socioeconomic gap (Coates, 2000). Regarding the effect of CO_2 emissions on the environment, some studies showed concern for environmental pollution by shifting centers of production to developing countries, where environmental regulations are not strict (Stiglitz, 2002); and other studies stated the possibility of reduced environmental load because of support for the transfer of environmental technology (Organisation for Economic Co-operation and Development, 2008b). It is known that CO_2 emissions in a country are affected by other countries through trades of goods and services. Developed countries generally import CO_2-intensive goods and export non-CO_2-intensive goods, and therefore, they are causing CO_2 emissions in developing countries.

In case of U.K., for example, Helm and colleagues (2007) calculated the CO_2 emissions caused by production of goods and services imported to the country and exported from the country, and they concluded that whereas CO_2 emissions in U.K. in 2006 were approximately 600 million tons-CO_2 (let us call them "statistical CO_2 emissions"), the net CO_2 emissions of the country that are caused by consumptions in the country and are taking trades into account (let us call them as "consumption-based CO_2 emissions") were 50% more than the statistical CO_2 emissions. Similarly, Hoshino and colleagues (2010) estimated that, in case of Japan, the consumption-based CO_2 emissions are 40% higher than the statistical CO_2 emissions; and that, in case of China, which manufactures goods to export to other countries, the consumption-based CO_2 emissions account for 27% lower than the statistical CO_2 emissions.

However, few studies discuss the effect of dynamic promotion of international division of labor on CO_2. In a quantitative discussion, Chung and Gillespie (1998) stated that international division of labor would promote economic development and therefore increase CO_2. Miozzo and colleagues (2005) provided a deeper discussion of their investigation regarding the effect of international division of labor on CO_2 in textile and chemical industries. They surveyed the changes of CO_2 emission and the value of production in Western Europe, the U.S., and China over the past 30 years. They showed that the total value of production of the three regions is increasing, the amounts of production and CO_2 emissions are shifting from Western Europe and U.S. to China, the amount of CO_2 is unchanged, and the CO_2 per value of production is decreasing. They stated the importance of introducing new technology and improved CO_2 efficiency in Western Europe, the U.S., and China in the future.

Data Review 1: Energy Consumption in Industrial Sectors of OECD and Non-OECD Countries

To assess the total CO_2 impact of CO_2 emission exports, first, the data for energy consumption change in the industrial sectors of Organisation for Economic Co-operation and Development (OECD) member countries (developed countries: 30 member countries as of 2009) and non-OECD member countries (developing countries) over 35 years are surveyed. Figure 11.1 shows the trends from 1971 to 2006, demonstrating that, while the energy consumption of OECD countries' industrial sectors decreased by 1.2% during those 35 years, that of non-OECD countries increased by 150%. In OECD member countries, although industrial energy consumption was unchanged

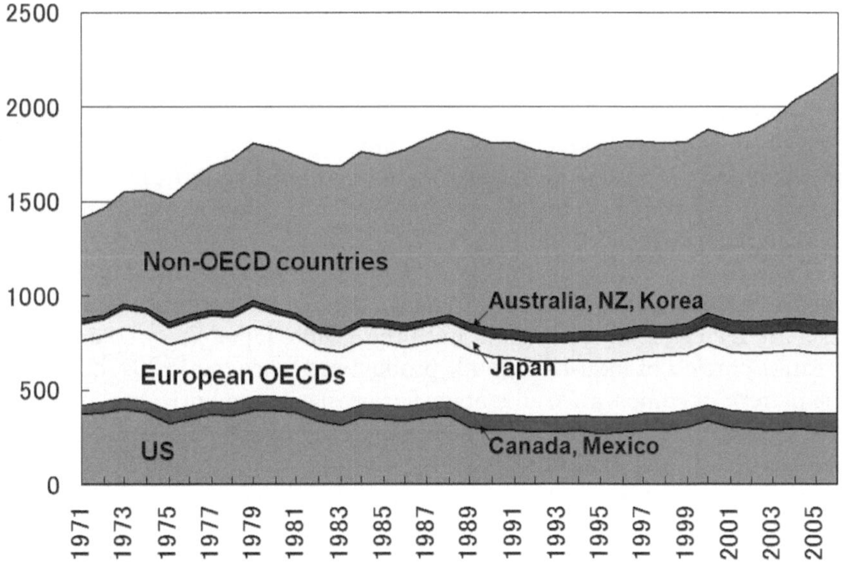

Figure 11.1 Trends of final energy consumption in industrial sectors of OECD and non-OECD countries (1971–2006).
Unit: Oil equivalent conversion, million tons
Source: Organisation for Economic Co-operation and Development, 2008c.

for Japan, it considerably increased for Korea and Mexico because of industrial development and resultant CO_2 imports. (Korea's emissions increased 7.6 times during 1971–2006; Mexico's emissions increased 2.3 times during the same period.) In contrast, the U.S. and many Western European nations' CO_2 emissions decreased considerably (24% decrease for the U.S. during the same period; 37% decrease for the U.K.). The trends shown in figure 11.1 probably result from the following causes.

- Production has been migrating from developed to developing countries.
- Production energy efficiency in developed countries has increased during this period.
- Production energy efficiency in developing countries is probably worse than that in developed countries, contributing at least partially to the increase in total energy consumption in the global industrial sectors.
- Thus, it is estimated that CO_2 exports (migration of production and services from developed to developing countries) will increase the global CO_2 emissions, at least in the short term.

Data Review 2: CO_2 Emissions in Specific Sectors in East Asian Countries

Figure 11.1 shows that whereas in developed countries, energy consumption in the industrial sector has remained unchanged in the 35-year period, it has increased significantly in developing countries. This increase implies the possibility that CO_2 exports have negative effects. However, to assess the amount of energy increase in developing countries that can be attributed to transfers of production of CO_2 emissions and the amount that can be attributed to increase in production in developing countries irrespective of production emissions transfers from developed countries, it is necessary to analyze more detailed data.

To assess the total impact of the CO_2 emission exports, the data of production trends and CO_2 emission trends in specific industrial sectors in East Asia are reviewed. In this region, until the 1980s, Japan accounted for the greatest portion of industrial goods production. Since the 1990s, however, the pattern of comparative advantage has changed, and horizontal international division of labor has increased, with a significant proportion of final products being manufactured in Korea, Taiwan, and other Southeast Asian countries. In recent years, China has emerged as the center of production in the region. Today, international division of labor and trade among nations continues to increase. Figure 11.2 shows the intermediate industrial goods trade in the region in 1990 and 2003, demonstrating the increased international horizontal division of labor during this period.

The spread of ICTs has significantly contributed to the establishment of international division of labor. For example, growing use of ICT to control

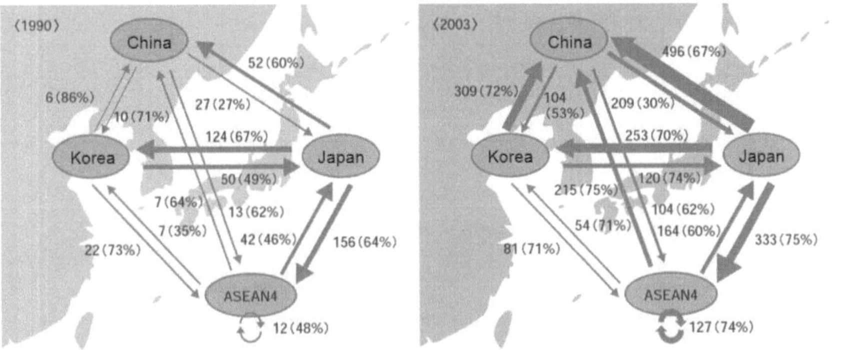

Figure 11.2 Trades of intermediate industrial goods in East Asia in 1990 and 2003. Numbers in () denote the share of intermediate goods trade in the total industrial goods trade.

Source: Ministry of Economy, Trade and Industry (METI), Ministry of Land, Infrastructure, Transport and Tourism (MILT), and Ministry of Education, Culture, Sports, Science & Technology (MEXT), Japan, 2006.

overseas operations has enabled multinational corporations to switch among possible locations and subcontractors for lower wages. ICT has also facilitated the development of industrial districts with networks of small, innovative, flexible, and high-quality product firms employing the latest technologies and professionals.

We selected two industrial sectors, textiles and electronics, for their changes in international division of labor over the last 20 years in the region. Tables 11.3 and 11.4 show the trends of production outputs and CO_2 emissions, respectively, in the textile sector in Japan, Korea, and China. Table 11.3 shows that the output in China increased rapidly, approximately three times during the period, whereas Japan's and Korea's paces are slower and actually declined. This data indicates that production has moved from Japan and Korea to China. China's CO_2 emissions increased greatly during the 1990s. Whereas the CO_2 per production output in China improved during the period, the energy efficiency level is still much lower than that of Japan and Korea. CO_2 emissions in China are more than 12 times those in Japan. The production of the same monetary value of textile products in China causes more than four times the CO_2 emissions than in Japan. Therefore, transferring textile goods production from Japan to China increases their total CO_2 emissions.

The electronics sector has grown remarkably since the 1990s, directly reflecting the ICT revolution. The production outputs and CO_2 emissions in Japan, Korea, and China are shown in tables 11.5 and 11.6, respectively. Table 11.5 indicates that the outputs of all three countries increased during the 1990s. In China, the production increased by a factor of more than 10 in that decade. Table 11.6 shows that CO_2 emissions in the countries

Table 11.3 Production Output of Textile Sectors (unit: 1,000 US$)

year	Japan	China	Korea
1990	107,439,724	78,400,088	34,821,297
1995	128,623,432	163,601,373	42,720,448
2000	73,476,048	208,902,146	39,945,263

Table 11.4 CO_2 Emissions in Textile Sectors (unit: million tons-CO_2)

year	Japan	China	Korea
1990	5.10	14.66	3.00
1995	3.49	22.41	2.96
2000	2.22	28.07	3.01

Table 11.5 Production Output of Electronics Sectors (unit: 1,000 US$))

year	Japan	China	Korea
1990	346,677,596	23,219,389	32,688,847
1995	537,032,050	76,707,954	73,743,978
2000	506,156,740	257,188,606	109,726,242

Table 11.6 CO_2 Emissions in Electronic Sectors (unit: million tons-CO_2)

year	Japan	China	Korea
1990	12.59	7.04	1.14
1995	11.87	16.30	2.62
2000	16.05	41.59	3.78

Source: Adapted from (IDE-JETRO, 1990; 1995; 2000) and the Organisation for Economic Co-operation and Development, 2008b.

also increased considerably, with the trend most significant in China, whose energy efficiency level is much lower than those of Japan and Korea. Whereas the production output of the electronics sector in China is about half that of Japan, China's CO_2 emissions are more than 2.5 times that of Japan. These data demonstrate that transferring electronics goods productions from Japan and Korea to China increased the total CO_2 emissions.

The data indicates that CO_2 emission exports increase total global CO_2 emissions, because CO_2 efficiency of developing countries, such as CO_2 intensity of electric generation, are usually worse than those in developed countries.

Data Review 3: CO_2 Export Impact Assessment Based on Product LCA and Electricity CO_2 Intensity Data

The data on the trends of production outputs and CO_2 emissions in the textile and electronics sectors in the 1990s in East Asian countries indicated the possibility that the transfer of production from Japan and Korea to China in the textile and electronics sectors and the low CO_2 production efficiency of China compared to Japan and Korea will lead to decreasing total CO_2 efficiency of production because of CO_2 exports. However, an issue that should be noted is that the goods produced in an industrial sector (e.g., textiles) in Japan and China might be different, although theoretically they may be within the same sector. Similarly, the features of goods produced in developed countries might be different from those produced in developing countries, although they may be in the same general category. In such cases, the CO_2 efficiency of an industrial sector cannot be compared among different countries. Therefore, in this section, we assess the effects of CO_2 export reflected in product LCA data and energy efficiency data in several countries and simulate the CO_2 emission impact when the products' production is offshored (transferred) from developed countries to developing countries. We consider automobiles, personal computers, and office work as examples and assess the impact of their CO_2 exports by referring to the data of CO_2 intensity of electricity in several countries.

Figures 11.3 and 11.4 show the LCA data of automobiles and personal computers, respectively, in Japan, and they show the CO_2 emissions of a product during the product life stages. For automobiles, CO_2 emissions of 1,850 kg-CO_2 are generated in the material production stage, 1,040 kg-CO_2 emissions in the component production and assembly stage, and 22,300 kg- CO_2 emissions at the usage stage (by gasoline consumption), that is, an automobile generates 26,000 kg-CO_2 emissions throughout its life cycle. A personal computer generates CO_2 emissions of 41 kg-CO_2 in the material production stage, 7 kg-CO_2 in the component production and assembly stage, and 33 kg-CO_2 in the usage stage, that is, total 84 kg-CO_2 throughout its product life cycle.

Figure 11.5 shows the CO_2 emission coefficients of electricity in different countries. The values of CO_2 emission coefficients of electricity are determined by the countries' electricity energy sources (e.g., CO_2 coefficients are high, or bad, in countries that have many coal-fired power plants but low, or good, in countries in that considerable amount of electricity is generated by renewable energy, such as hydraulic power, or that have many nuclear power plants) and by the efficiency of power plant facilities (e.g., CO_2 coefficients are low in countries with highly efficient coal-fired power plants but high in countries with a large number of inefficient coal-fired power plants). The CO_2 emission coefficient in Japan is 0.40 kg-CO_2/kWh. In Japan, approximately 30% of electricity in the year was generated by nuclear power plants, and the thermal power generation plants are highly efficient. On the other hand, in China and India, where coal-fired power plants comprise the majority (80% in China and 70% in India), the CO_2 coefficients are more than double that of Japan (0.86 kg-CO_2/kWh in China and 0.96 kg-CO_2/kWh in India).

In France, where nuclear power plants generate 80% of electricity and in Canada where hydraulic power comprises almost 60%, CO_2 coefficients are very low.

CO_2 emissions in the material production and component production and assembly stages in figures 11.3 and 11.4 are counted as industrial sector emissions. If we assume that all energy used in the production stage is electricity, then transfer of production from Japan to China doubles the

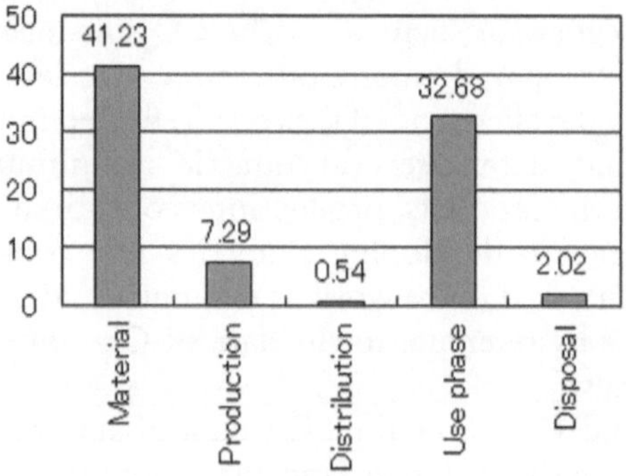

Figure 11.3 LCA CO_2 emissions of a desktop personal computer (unit: kg-CO_2).
Source: Japan Environmental Management Association for Industry: http://www.jemai.or.jp/JEMAI_DYNAMIC/data/current/prodobj-2309-pdf.pdf

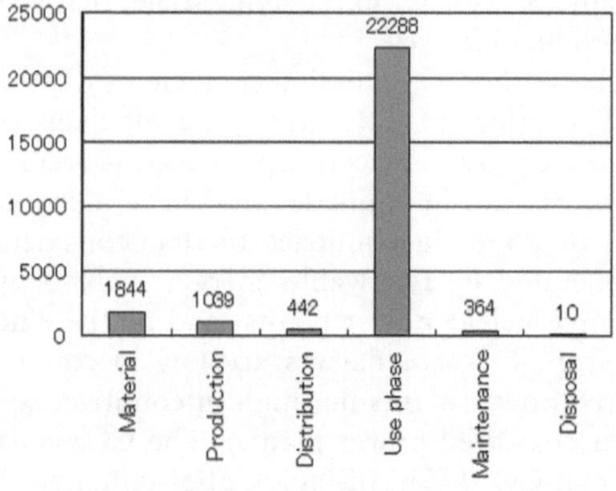

Figure 11.4 LCA CO_2 emissions of an automobile (unit: kg-CO_2).
Source: Japan Automobile Manufacturers Association: http://www.jama.or.jp/lib/jamagazine/199806/02.html

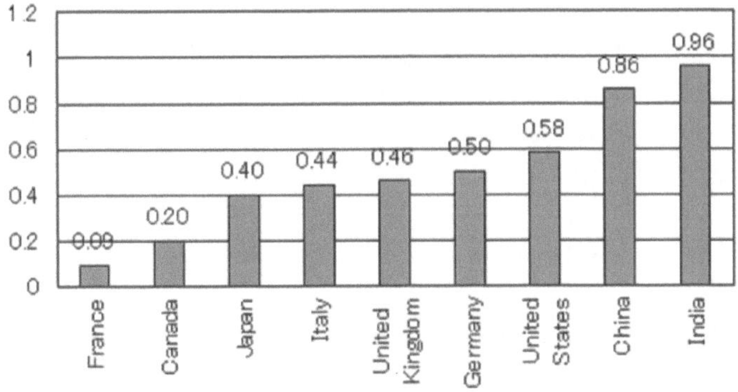

Figure 11.5 CO_2 emission coefficients of electricity.
Source: Organisation for Economic Co-operation and Development, 2008a.

CO_2 emissions of its production with respect to the CO_2 coefficients of electricity. However, material production processes that use energy other than electricity, such as those using steel and similar materials (e.g., automobiles and other products using steel and other heavy components), are technology- and capital-intensive. Thus, compared with the transfer of lighter product assemblies, it is more difficult to transfer such production to other countries. We compare the CO_2 emissions in the production stages for both automobile production and personal computer production, where the products are manufactured in Japan (figures 11.3 and 11.4) and where the products are manufactured in other countries. In the latter situation, we assume two cases:

1) CO_2 efficiency of production stages in a country is proportional to the CO_2 efficiency of electricity (figure 11.5) in the country.
2) Only the CO_2 efficiency in the component production and assembly stage during production is proportional to the CO_2 efficiency of electricity (figure 11.5), and the CO_2 efficiency in the material production stage is unchanged because it is assumed that material production is not transferred.

Figure 11.6 shows the results. If the production of automobiles is transferred from Japan to China, for example, under the former assumption, the CO_2 emissions of automobile production double. Under the latter assumption, the CO_2 emissions increase by a factor of 1.4. The CO_2 emissions are worse if the production is transferred to India.

Similarly, in the case of personal computers, transfer of production from Japan to China doubles the CO_2 emissions under the former assumption and increases it by a factor of 1.3 under the latter assumption.

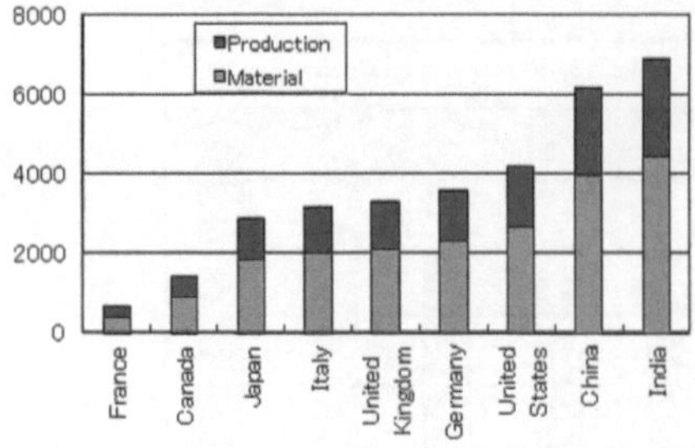

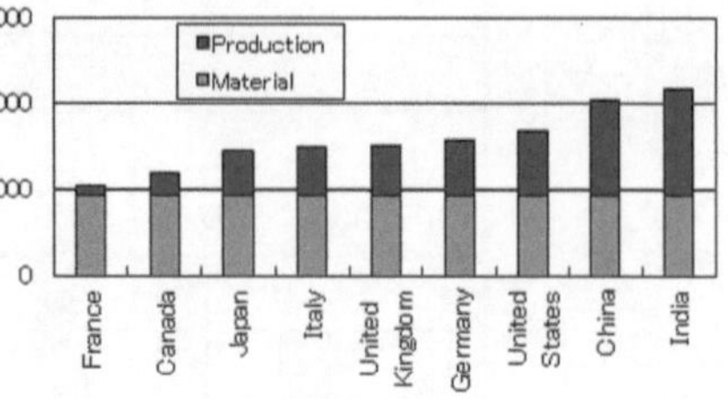

Figure 11.6 CO_2 emissions of an automobile in material and production stages in different countries.
(Left) Both material and production stages are transferred to the countries and CO_2 emission amounts are proportional to the CO_2 emission coefficient of electricity in the countries.
(Right) Production stage is transferred to the countries and CO_2 emission amounts are proportional to the CO_2 emission coefficient of electricity in the countries.

In addition, in the case of offshore outsourcing of office work, such as that of impersonal services (table 11.1), because most of the energy consumed in office work is electricity, the CO_2 impact of offshoring depends on the difference between the CO_2 coefficients of electricity in the outsourcing country and outsourced countries. Outsourcing to countries such as China and India, which have high CO_2 coefficients, increases the total CO_2 emissions.

SUMMARY: ICT'S NET EFFECT ON CO_2

ICT's effects on CO_2 emissions examined in this chapter are summarized in table 11.7. In the automobile manufacturing sector, for example, ICT's direct CO_2 effects include the following: (1) CO_2 increase due to the increased use of ICT equipment in the auto manufacturing sector, and (2) CO_2 reduction due to the expansion of CO_2-saving ICT applications, such as video conferencing and paperless offices in the auto manufacturing sector. ICT's indirect CO_2 effects are the same as those of transferring auto manufacturing from industrialized countries to developing countries, which are at least partially caused by ICT's effects on socioeconomics. Because of the high CO_2 coefficients in developing countries in general, the production migrations increase CO_2 emissions from auto manufacturing by 1.4 times to 2 times (figure 11.6).

The net impact is the total of direct and indirect effects. Referring to the results of the existing studies outlined in section 1 (table 11.1), the amount of direct impact is estimated as several percent of CO_2 emissions by the sector, whereas the amount of indirect impact is several 10s percent of the emissions by the sector. Thus, indirect impact is larger than direct impact, and the net impact is estimated as negative or estimated to increase CO_2

Table 11.7 Summary of ICT's Effects on CO_2

sector	ICT's direct effects on CO_2	ICT's indirect effects on CO_2	ICT's net total effect on CO_2
Automobile manufacturing sector	– CO_2 increase due to increased use of ICT equipment by auto manufacturing sector – CO2 reduction due to expansion of CO2-saving ICT applications (TV conference, paperless office, etc.) by auto manufacturing sector	– CO_2 increase due to migration of production from industrialized countries (low CO_2 coefficient) to developing countries (high CO_2 coefficient) (figure 11.5) 1.4–2 times CO_2 emissions per auto production (figure 11.6)	
Electronics sector	– CO_2 increase due to increased use of ICT equipment by electronics sector – CO_2 increase due to increased use of ICT equipment production (tables 11.5 and 11.6) – CO_2 reduction due to expansion of CO_2-saving ICT applications (TV conference, paperless office, etc.) by electronics sector	– CO_2 increase due to migration of production from industrialized countries (low CO_2 coefficient) to developing countries (high CO_2 coefficient) (figure 11.5)1.2–2 times CO_2 emissions per production (section 3.4)	
Entire society	– CO_2 increase due to increased use of ICT equipment 1%–3% of total CO_2 emissions (table 11.1) – CO_2 reduction due to expansion of CO_2-saving ICT applications~5% of total CO_2 emissions at potential (table 11.1) – In total, a potential of ~3% reduction in total CO_2 (table 11.1)	– CO_2 increase due to migration of production from industrialized countries (low CO_2 coefficient) to developing countries (high CO_2 coefficient) (figure 11.1 and other data)Likely to be a large amount. A quantitative figure is not estimated.	Sum of direct and indirect effects Likely to be negative (CO_2 increasing effects) in total

emissions in total. Similarly, direct and indirect effects of the electronics sector are assessed in previous sections. In society overall, the estimates of direct impact in previous studies were reviewed in section 1 (table 11.1). According to the estimates, the amount of CO_2 increase is estimated to be 1% to 3%, whereas that of CO_2 reduction is estimated to be up to 5%. Although the amount of indirect effects is not quantitatively estimated, as discussed in section 3, based on the observed effects of production migration from industrialized countries to developing countries where CO_2 coefficients are generally higher, it is likely that they experience considerably increasing CO_2 emissions, at least in the short term. Thus, the net CO_2 impact is also estimated as the combined effects of increasing CO_2.

CONCLUSION

In this chapter, ICT's impact on CO_2 emissions was explored to exemplify the assessment of indirect and unintended effects of innovation. Because ICT is one of the key factors enabling globalization today, assessing "CO_2 effects of globalization" is necessary to analyze ICT's net impact on CO_2 emissions. The effects were analyzed by viewing three types of data. The results indicated that globalization is likely to have yielded and to further yield the increase in global CO_2 emissions.

Recently, assessments of the socioeconomic effects of innovation have been attempted actively. Behind the trend are the large expectations for innovation by industry, governments, and societies. In addition, many governments today try to find methods of effective decision making on their innovation policy and their research and development investments based on the impact and benefit assessments of the investments for innovation. The attempts are sometimes referred to as "science of science policy." However, in most cases of these attempts, the direct and positive effects of innovation, especially in terms of economics, are highlighted, and the indirect effects and possibilities of unintended consequences are rarely considered. As shown in this chapter, however, innovation's indirect effects are often larger than direct ones. Accordingly, the assessment of not only direct effects but also indirect effects of innovation is necessary. Development of methods for accessing innovation's indirect effects is an important task.

Through the example shown in this chapter, it is indicated that to assess the indirect impact of innovation, it is first necessary to consider a wide range of socioeconomic impacts of a particular innovation. In this chapter's example, ICT's effects of promoting globalization were focused to assess ICT's impact on CO_2 emissions. It is necessary to take wide variety of socioeconomic effects into account and assess the multiplied impact of those effects. Second, because indirect effects of innovation are wide and complex, data assessment in several levels is effective. In this chapter, data on energy consumption in the industry sector in the world, CO_2 emissions in selected industrial sectors in selected countries, and CO_2 analysis in specific product levels were assessed. Relying on a single type of data has the risk of making us lose the holistic view. To get the full picture of the impact of innovation, it is not enough to just count the direct effects. One must also add indirect effects and unintended consequences.

REFERENCES

Aird, C. L., & Sappenfield, D. (2009). IT the 'enabler' of global outsourcing. *Financial Executive*, 25, 62–63.

Baldwin, C. Y., & Clark, K. B. (2000). *Design rules, Volume 1: The power of modularity*. Cambridge, MA: MIT Press.

Blinder, A. (2006). Offshoring: The next industrial revolution? *Foreign Affairs*, 85, 113–128.

Caudill, R., Luo, Y., Wirojanagud, P., & Zhou, M. (2000). *A lifecycle study of the impact of e-commerce on electronic products*. San Francisco: Proceedings of International Symposium on Electronics and the Environment.

Chung, C., & Gillespie, B. (1998). Globalisation and the environment: New challenges for the public and private sectors. In Organisation for Economic Co-operation and Development (Ed.), *Globalisation and the environment: Perspective from OECD and dynamic non-member economics*. Paris: Organisation for Economic Co-operation and Development.

Coates, D. (2000). *Models of capitalism: Growth and stagnation in the modern era*. Cambridge: Polity Press.

Friedman, T. (2005). *The world is flat*. New York: Farrar, Straus and Giroux.

Friedman, T. (2008). *Hot, flat, and crowded*. New York: Farrar, Straus and Giroux.

Gartner Research. (2008). *Green IT: A new industry shockwave* (PowerPoint presentation).

Helm, D., Smale, R., & Phillips, J. (2007). *The UK's climate change record, too good to be true*? Retrieved from http://globalisierungsklima.files.wordpress.com/2007/12/carbon_record_2007.pdf (accessed 31 July 2011)

Hoshino, Y., Sugiyama, T., & Ueno, T. (2010). International comparison of trade embodied CO2 emissions. *Journal of Japan Society of Energy & Resources*, 31, 8–14. (Japanese)

Institute of Developing Economies, Japan External Trade Organization (IDE-JETRO). (1990, 1995, 2000). *Asian international input-output tables*. Tokyo: IDE-JETRO.

Kawamoto, K., Koomey, J., Nordman, B., Brown, R., Piette, M., Ting, M., & Meier, A. (2002). Electricity used by office equipment and network equipment in the U.S. *Energy—The International Journal*, 27, 255–269.

Matsumoto, M., Hamano, J., Tamura, T., & Iguchi, H. (2007). Impacts of ubiquitous technology advances on energy consumption in Japan. *Electrical Engineering in Japan*, 161, 22–30.

Ministry of Economy, Trade and Industry (METI), Ministry of Land, Infrastructure, Transport and Tourism (MILT), and Ministry of Education, Culture, Sports, Science & Technology (MEXT), Japan. (2006). *White paper of manufacturing* (Japanese). Tokyo: Gyosei.

Ministry of Internal Affairs and Communications. (2008). *Report of committee on ICT policy to combat global warming* (Japanese). Japan Ministry of Internal Affairs and Communications. Retrieved February 28, 2011, from http://www.soumu.go.jp/main_sosiki/joho_tsusin/policyreports/chousa/ict_globalwarming/pdf/0804_h1.pdf

Miozzo, M., Dewick, P., & Green, K. (2005). Globalisation and the environment: The long-term effects of technology on the international division of labour and energy demand. *Futures, 5*, 521–546.

Nakamura, M., Takahashi, K., Nishi, S., Kamekura, M., & Takeshima, M. (2000, October). *Lifecycle inventory study of teleconference system*. Tsukuba, Japan: Proceedings of the 4th International Conference on EcoBalance.

Organisation for Economic Co-operation and Development. (2008a). *CO_2 emissions from fuel combustion*. Paris: Author.

Organisation for Economic Co-operation and Development. (2008b). *Environmental outlook to 2030*. Paris: Author.

Organisation for Economic Co-operation and Development. (2008c). *Energy balances of OECD countries*. Paris: Author.

Origuchi, T., Nishi, S., & Fujimoto, J. (2007, December). *Estimation of reducing CO2 emissions through introduction of ICT in manufacturing and distribution sectors*. Tokyo: Proceedings of the 5th International Symposium on Environmentally Conscious Design and Inverse Manufacturing (EcoDesign 2007).

Rodriguez, C., Wunnik, C., Sancho, L., Burgelman, J., & Desruell, P. (Eds.). (2004). *The future impact of ICTs on environmental sustainability*. (European Commission Technical Report EUR 21384 EN). Seville: European Commission Joint Research Centre.

Romm, J., Herrmann, S., & Rosenfeld, A. (1999, December). *The Internet economy and global warming*. Arlington, VA: Center for Energy and Climate Solutions.

Roth, K. (2005, October). *The potential to reduce information and communication: Technology (ICT) energy consumption* (PowerPoint presentation). Conference Special CEPE-Colloquium, Zurich.

Smith, T. (2000). *Technology and capital in the age of lean production: A Marxian critique of the 'new economy.'* New York: State University of New York Press.

Stiglitz, J. (2002). *Globalisation and its discontents*. London: Penguin.

Takase, K., & Murota, Y. (2004). The impact of IT investment on energy: Japan and US comparison in 2010. *Energy Policy, 32*, 1291–1301.

Toffel, M., & Harvath, A. (2004). Environmental implications of wireless technologies: News delivery and business meetings. *Environmental Science and Technology, 38*, 2961–2970.

Ueda, H., Nakazawa, K., & Hashitani, T. (2004, October). *Feasibility of reducing CO_2 emissions by modal shift in commuting*. Proceedings of the 6th International Conference on EcoBalance, Tsukuba, Japan.

Williams, E., & Tagami, T. (2002). Energy use in sales and distribution via B2C E-commerce and conventional retail: A case study of the Japanese book sector. *Journal of Industrial Ecology, 6*, 99–114.

12 Challenging the Innovation Paradigm

Conclusions, Practical Implications, and Future Research

Karl-Erik Sveiby, Pernilla Gripenberg, and Beata Segercrantz

Human innovativeness is a very valuable capacity and that has probably been recognized since the dawn of humanity. When we engage in innovation, we introduce change, something that makes a difference in the established order. It is this powerful feature of innovation that has made it part of political and religious discourses as long as we have written records, that is, in the West since antiquity. Godin shows in chapter 3 how the concept of innovation has always been exploited by the rulers of the day. Historically innovation was promoted as 'bad' and used as a pejorative by the church and kings because it threatened the(ir) established order. Today, the concept is still exploited to serve a political agenda for the establishment of our day, an amorphous amalgamation of commercial interests and national politics, and in an ironic twist, innovation is now promoted as 'good.' We are told that there is a global competition between companies and nations in which only the 'best' innovators will prevail. Innovation has become an inevitable imperative; 'we' *must* innovate. More, better, faster—or else.

The purpose of this book is to challenge that view. We as editors of this book are not anti-innovation—on the contrary. We believe that innovativeness is an inherently human faculty and as such it is 'good.' Our concern is the way in which the result—innovation, is understood, researched, and exploited. Fougère and Harding track in chapter 2 the currently prevailing *managerial view* of innovation to the 1960s, with the publication of the Charpie Report and the *Oslo Manuals*, which define innovation as changes that can be counted. Hence, what cannot be counted is not innovation, and it becomes impossible to speak of failed innovations—an innovation that fails is not an innovation. Innovation research has been highly instrumental in this process by providing empirical evidence that innovating firms achieve higher economic performance, thereby allowing policy makers and business leaders to legitimatize innovation policies and public investments in research and development (R&D) and innovation.

Sveiby, Gripenberg and Segercrantz show in chapter 4 that research with noneconomic research perspectives and research counter to the paradigm, such as studies of negative effects and undesirable consequences of innovation for other stakeholders than the innovating firms and change agents, have become marginalized in the field of 'innovation studies.' Fougère and Harding highlight how the type of deeper longitudinal sociological studies exemplified by Everett Rogers, no longer constitute or belong to any of the dominant 'cognitive communities' within 'innovation studies.' Lindell draws in chapter 8 attention to the theoretical models for new product development. They have, at best only minimally, considered non-beneficial effects of innovation on society.

Studies, which take a broader stakeholder view and discuss negative social and environmental impacts, are still made in research streams under labels of technology assessment (TA) and social shaping of technology (SST). Ecological effects of innovation and side effects of medicines are scientific fields in their own right. None of those fields have an impact on 'innovation studies,' however.

Innovation studies seem to have implicitly isolated itself by developing terminologies and applying theories from economics in particular. It is a separation of discourses with contentious implications for at least two reasons. First, beneficiaries of the desirable economic effects and recipients of the undesirable consequences tend to be different groups. Contributors to this volume demonstrate that innovation, apart from the intended beneficial economic effects enjoyed by the innovators and change agents; also has adverse—even highly negative—unintended effects and consequences for other stakeholders in society.

Second, there are two ways to increase the net effect of innovation for individual firms as well as for society as a whole: to increase the desirable consequences and to reduce the undesirable ones. This may seem a trivial observation, but it apparently needs pointing out. A one-sided focus on direct desirable effects of innovation misses many opportunities to reduce the indirect undesirable consequences. They can be considerable as authors show in this volume, and due to systemic effects, they reduce the net benefits for individual employees, customers, as well as for society as a whole (see chapters 6 and 11 in this volume)—even the innovators' own organizations (see chapters 6, 9, and 10 in this volume).

Particularly worrying is that the undesirable consequences of innovation might be on the rise globally in relative terms, that is, not only due to increased innovation volume. Authors identify two possible causes in this volume. One is that decision makers have embraced the 'speed imperative,' the current portrayal of innovation as a competition, and they are continuously trying to accelerate the pace. Acceleration may help developing regions to advance and become more prosperous quicker, and to address social challenges, but as Hasu,Leitner, Varblane, Solitander show in chapter 5, the speed imperative also reduces reflexive thinking. The financial innovation case gives support to their view. Sveiby shows in chapter 6 how creativity went down when innovation volume went up in the

2000s. Simultaneously, when innovation intensified, otherwise competent actors unwittingly made dangerous judgment errors, that is, risk went up. Speeding up became dumbing down. Maybe there exists a level of intensity beyond which innovation becomes counterproductive? It certainly appears to be the case in financial innovation. Shouldn't we try to figure it out before we unwittingly accelerate innovation beyond the 'speed limit'?

The other driver behind increasing levels of undesirable consequences is the systemic nature of innovation. Consequences of errors spread rapidly on a global scale these days, enabled by ICT innovations. The financial services world was already in the late 1990s so highly interconnected that the decisions of one individual portfolio manager in one single hedge fund (Long Term Capital Management) threatened the foundations of the largest U.S. investment banks and thereby indirectly the global financial market. This executive based his decisions on a financial innovation, a 'foolproof' theoretical model for arbitrage trading, developed in-house. ICT innovations also enable companies to shift production from Western countries to Asia, a routine decision having been made thousands of times by thousands of managers. Each one of them thereby unwittingly 'exports' CO_2 emissions and contributes indirectly to a yearly net increase in CO_2 emissions globally, as Matsumoto and Kawajiri illustrate in chapter 11.

Thirdly, we do not know what systemic effects society-wide acceleration of innovation in many fields simultaneously may cause.

In view of the (possibly) increasing systemic effects, it is a concern that the close connection between innovation and its undesirable consequences 'gets lost' in the public debate, as well as in policy. The separation of research discourses appears to be mirrored by compartmentalization on policy level. With some exaggeration, the ministers of industry and economy are guided by innovation studies and may become 'ministers for desirable consequences of innovation,' whereas the heads of the health, social and environment departments can be called 'ministers for the undesirable consequences of innovation.'

PRACTICAL IMPLICATIONS

J. R. R. Tolkien warns: 'It does not do to leave a live dragon out of your calculations, if you live near him.' (cited by Koehn, D. 2010 p.3). For the practitioner, unintended consequences of innovation are akin to Tolkien's dragon. The trouble for innovators and other practitioners involved with product development is that, at any point in time, many or most consequences of innovation are possible to know only with hindsight, and then often only with a hindsight bias. Some implications for practitioners to bring the dragon into their calculations follow.

For Innovation Executives

The dragon is at risk of 'awakening' metaphorically at any time during the innovation process—particularly when the innovation leaves the comparatively safe in-house environment. Rarely, the consequences are as public as in the financial innovation case. More often than not, the dragon's destructive activities remain deceptively hidden for some time while the innovator's/change agent's/policy maker's views are dimmed by obsolete measuring tools and excessive amounts of irrelevant data. More information of the same kind does not improve assessments, as several examples from the financial innovation case in chapter 6 show. Hence, as suggested by Leitner in chapter 7, weak signals, such as the opinions of silenced voices or trends in the fringes, can be of more value. Those types of data are, however, too easily dismissed, ridiculed, or ignored when change agents push the mainstream agenda or when product developers have executives breathing down their necks with demands to speed up the process.

Therefore, contributions by independent and contrarian thinkers should be encouraged in corporate R&D. In the financial innovation case, critical inputs by scholars would have balanced the inability of industry experts to question the fundamental errors of the financial products they are developing. Related to this is the recommendation in chapter 5 to take reflective breaks; intermittently looking back, probing the fundamental issues and discussing them. Such breaks constitute the first step in double-loop learning and should be included in NPD processes also after the launch phase, as Lindell recommends. Particularly fruitful moments to learn occur after critical events; the road to the financial meltdown 2008 was quite long and littered by lost opportunities to do so.

The belief that innovation is essential for profitability, survival, and success is repeated so frequently that it has become an axiom. As we can show in this volume, however, the belief is false. Less innovative products can be more profitable than highly innovative ones (banking) and imitation can be more efficient than innovation (franchising) are two examples from the literature search discussed in chapter 4.

Is it good practice to launch new products that have not been tested outside the laboratory? Most industry executives would say no. However, as Sveiby shows in chapter 6, such behavior is normal practice in the financial industry. Chapter 5 suggests that companies should take a long-term perspective on their strategies and critically ask themselves how to build up a sustaining innovation culture and capacity. Hasu and colleagues provide a good practical example of such a framework in chapter 5: Jacques Ellul's '76 Reasonable Questions to ask about any innovation.' Those questions are the closest we come to a discussion of ethics in this book. There is no chapter on the ethics of innovation, but it is not for want of trying. Perhaps symptomatic of this research field, all efforts to produce a high-quality chapter on the topic have failed for various reasons. This is deplorable, because most of the issues highlighted in this book contain practical dilemmas with ethical ramifications. Innovation executives could take example from the lively debate among medical scientists and practitioners about ethical issues in relation to medical innovations.

For Managers and HR Practitioners

Chapter 10 highlights the complexity of consequences for employees in a case of organizational innovation. Cañibano and colleagues note that the main direct effect of such innovations tend to be one of work intensification, and that for the managers in their case company, there is no simple answer as how to deal with the indirect effects of it. The case goes to the ethical core of what it means to consider consequences before one introduces an innovation. The practitioner needs to ask uncomfortable questions such as those in chapter 5. Their chapter emphasizes the importance for managers and human resources practitioners to consider the total impact of organizational innovations have on employees.

For Product Developers

The contributors to this volume put forward other practical suggestions. One suggested in chapter 5 is to apply TA and SST approaches more widely in product development models to include ecological and ethical issues. Chapter 1 and 6 suggest that a stakeholder view is a useful perspective in theory building as well as in practice. Lindell suggests that innovation models should be more connected to their context, both internally and externally, and that the core development process is extended to include the implementation phase of the innovation, as well as diffusion, growth, and product retirement phases. A case in point is the financial industry, where improved testing procedures during the implementation phase might have avoided some of the most blatant mistakes during the lead-up to the general financial crisis. The lack of risk control inherent in innovation in the financial industry has become glaringly obvious; boards in other industries also need to take heed. However, as Segercrantz notes in chapter 9, when formalization and control of innovation work turn creative ICT developers to compliant resources, it can be followed by dis-identification, reduced creativity, and loss of overview of the product they are developing. Unintended indirect consequences of such effects have not been studied.

A common belief is that 'the only way to reduce negative consequences of innovation is to continue innovating' as one of the authors in the literature review in chapter 4 states. Of course humanity needs more innovation to take us out of our global challenges today. However, many of our challenges today are direct effects and indirect consequences of older innovations. In the financial industry case, innovation improved efficiency and accelerated new varieties of the existing design, but it did not address the fundamental flaws in the products. The effect was merely to accelerate the financial industry's descent into its death spiral. Changes in behaviors and lifestyle can be more effective than innovation, for instance, in reducing CO_2 emissions (often caused by previous innovations) than more technical innovation.

For Educators

Contrarian thinking, critical reflection, corporate social responsibility, and business ethics are on the curriculum of management education in most business schools. For some reason, many if not most textbooks on innovation are primary examples of the pro-innovation bias, and management and business education plays an important role in spreading biased innovation models. Innovation textbooks should help students to discover biases in contemporary innovation research and literature as Lindell suggests.

FUTURE RESEARCH

Godin compares in his chapter the lack of critical debate about the concept of innovation with theology, citing Hans Blumenberg (1979), who once said that the "basic embarrassment of every theology" is "to speak about God constantly without having the right to permit itself to say anything about him." The lack of research into negative effects and unintended consequences of innovation has a positive element—at least for researchers: the field is practically untouched and in dire need of innovation. Rogers (1983) suggested studies with an 'anti-innovation bias' to balance the prevailing pro-innovation bias. None of the contributors in this book subscribes to that view. The issue for us is rather one of concern about the future for society if the current innovation paradigm is allowed to prevail. Policy makers, change agents, and other practitioners currently receive far too little guidance from innovation research on how to consider the less beneficial consequences of innovation. They thereby unwittingly reduce the net positive effect of innovation and contribute to unnecessary suffering for employees and consumers, not to mention the innocent bystanders, all of which reduce the net value of innovation for society. Both theoretical and methodological developments are required. There is a need for theoretical development, through which we can learn more about the impact of innovations in the social world. Innovations affect society, sometimes adversely (yet unintentionally), but the effects have been marginalized by innovation scholars, as they do not fall within the economic domain, but rather emphasize ecological, social, moral, and other non-economical values. Not much has happened since Rogers's socially grounded approach; one suggestion for amendment due to the built-in pro-innovation bias in diffusion theory is proposed by chapter 4 in this volume. Fougère and Harding advocate a more thorough and critical rehabilitation of diffusion theory; it might allow us to identify not only undesirable consequences of innovation, but also to help us better understand the changes that are attributed to 'innovation.' They see this discussion as essential because innovation has become crucial to the identity through which the West knows itself. Chapter 5 contributes to this line of thought by encouraging researchers to go beyond the current innovation paradigm and incorporate theories and methods from the TA and SST fields and give examples of methods, which could elaborate and extend the research agenda of traditional innovation research.

The direct undesirable effects of organizational innovations on employees and consumers show up most frequently in literature searches, particularly introductions of ICT. Is this because ICT is particularly prone to "mindless" implementation, or is it because there is an established stream of research in TA and SST literature for studying them? What might we find if there was a corresponding stream of research in 'innovation studies'?

Research highlighting the *indirect* undesirable effects of innovations on society, social structure, and bystanders is rare indeed. Still, they do exist: Chapter 4 cites a prescient 2001 study warning for the indirect consequences of financial innovation, and Matsumoto and Kawajiri contribute a study of indirect negative effects of ICT innovation on CO_2 emissions in this volume. Indirect effects are complex and multilevel, and they are therefore particularly difficult to study with quantitative methods, especially if they rely on data collected with predetermined definitions, such as the *Oslo Manual*. While such studies have longitudinal value, they lock scholars into path dependency and they risk becoming self-referring and 'routinized.' An example of such behavior is noted in chapter 6, which highlights that critical and reflective texts by academic experts are few and far between in the financial field. Non-reflective innovation research blinded by perspective or ideology is not only poor science; it also contributes to legitimize dangerous products and unethical behaviors. New and different kinds of data need to be developed that take into account a wider range of socioeconomic impacts of an innovation. There is, however, little in terms of empirical evidence and cases about the implications of acceleration. We know very little about 'slow innovation,' and we need much more research to understand how and in which contexts innovation can be decelerated successfully or even avoided entirely. Meanwhile, decision makers should be more careful when introducing policies that may accelerate innovation in an industry, organization, or society.

The most important point therefore, is to free research as much as is humanly possible from implicit blinders, be they the prevailing paradigm, dominant axiomatic perspectives, or other factors that limit the competence of those involved. As the chapters by Leitner, Lindell, Sveiby, and Matsumoto and Kawajiri demonstrate in this volume, existing and publicly available data can be as good as new if the research questions are new and come from a different perspective. We also need 'deep' research. Chapters 2, 3, and 5 show the value of critical discursive analysis in revealing hidden patterns in extant literature, whereas chapters 9 and 10 display the depth and detail possible to reach with primary data.

CONCLUSION

There is no doubt that we, as humanity, need more innovation to improve life on this planet. To accomplish this huge task, we will need the combined innovativeness of all humanity. We also need to make innovation more effective, and a good starting point is to learn from old mistakes. An unavoidable observation is that many of the economical, ecological, and social challenges confronting us on the planet today are caused by direct effects and indirect consequences of previous innovations. The contribution of this book is to bring attention to the commercial waste, policy ineffectiveness, and human suffering caused by the way corporations have executed and policy makers have regulated innovation. We need reflective innovation—not reckless, mindful innovation—not mindless, effective innovation not only efficient. The good news is that there exist huge opportunities to increase the net benefits of innovation for society by learning from previous mistakes inherent in existing innovations. For this, we need to change the way innovation is understood on all levels in society. Scholars carry a particularly important responsibility in realizing that goal by supporting decision makers with good unbiased innovation research. To achieve this, innovation research needs to free itself from its paradigmatic isolation from reality and head toward a more valid view on innovation. By highlighting the issues involved and suggesting methods and areas for further study, the contributors and editors of this book have tried to demonstrate the value and the relevance in doing so.

REFERENCES

Koehn, D. (2010). Living with the dragon: Thinking and acting ethically in a world of unintended consequences. New York: Routledge.

Blumenberg H. (1997). Prospect for a theory of nonconceptuality. In *Shipwreck with spectator: Paradigm of a metaphor for existence.* Cambridge, MA: MIT Press. (Original work published 1979)

Rogers, E. M. (1983). *Diffusion of innovations* (3rd ed.). New York: Free Press.

Contributors

Oihana Basilio is a PhD candidate and teaching assistant of External Sector at the Department of Economic Structure and Economy of Development at the Autonomous University of Madrid (UAM). She holds a master's degree in Economics and Management of Innovation and a master's degree from the European Inter-University Association on Society, Science and Technology (ESST). She has presented her work at various international conferences.

Almudena Cañibano is a PhD candidate of the Department of Management at the London School of Economics. Her research focuses on the analysis of human resource management practices and flexible work arrangements and their effects on employee well-being. She is a teaching assistant at the London School of Economics and lectures in Centro de Estudios Garrigues (Madrid) and The Strategy Academy (Kolkata, India). She has presented her work at many international conferences and has coauthored a book chapter for the *Oxford Handbook of Participation in Organizations* (2010).

Martin Fougère is assistant professor in politics and business at Hanken School of Economics in Helsinki. His main research interests are in business and society, corporate responsibility, critical management studies, and governmentality. His works on these and other topics have been published in a number of international journals and books. He has written critical studies of powerful managerial discourses—in particular, corporate responsibility and creativity—that he sees as related in several ways to a 'pro-innovation bias.'

Benoît Godin is a professor at Institut National de la Recherche Scientifique (Montreal). He holds a DPhil in science policy from Sussex University (Brighton, UK). He has written extensively on science policy and statistics. Since 2000, he has conducted a project on the history of science and technology statistics from which two books have been published: *Measurement and Statistics on S&T: 1920 to the Present*, 2005, and

La science sous observation: cent ans de measures sur les scientifiques, 1906–2006, 2005. More recently, he has launched a large-scale project on the intellectual history of innovation as a category, from the Renaissance to the present day.

Pernilla Gripenberg, PhD, is a researcher at the Department of Management and Organization at the Hanken School of Economics in Helsinki, where she also has been working as a lecturer for many years. Her research interest is in human-ICT relationships and the sociocultural effects of ICT use on everyday life in and across various contexts (such as work and home). She has published her work in books and journals, such as *The Information Society* and *Information Technology and People.*

Nancy Harding is a professor of Organisation Theory at Bradford University School of Management. Her research focus is on working lives, and she is particularly interested in why work is so dispiriting and so oppressive for many people in the West. She has published papers in international journals, she has written sole-authored and coauthored books, and she has contributed to edited books as author and coeditor. Her preferred mode of publication is the book, and she laments the decline in importance of the book in organization studies in general and critical organization studies in particular.

Mervi Hasu PhD (Educ.) is a specialized researcher at the Finnish Institute of Occupational Health (FIOH). Her research is focused on social innovation in various private and public sector service contexts, focusing on *employee-driven innovation* and *well-being in innovation.* She has worked at the Technical Research Centre of Finland (VTT) and at the University of Helsinki. She has published articles on the implementation of innovation and on interaction and learning in the innovation process. She has also carried out research in various industrial and business contexts, ranging from the traditional machinery industry and its subcontractor networks to science-based spin-off firms and their user communities.

Kotaro Kawajiri is a research scientist at the Research Institute of Science for Safety and Sustainability at the National Institute of Advanced Industrial Science and Technology (AIST) in Japan and a visiting scientist at Massachusetts Institute of Technology (MIT) in the US. His research focuses on life cycle assessment (LCA), supply chain management, and environmentally benign manufacturing. He is the author/editor of three books and several academic articles.

Karl-Heinz Leitner is a senior scientist in the Department of Foresight and Policy Development at the Austrian Institute of Technology. His main research interests cover innovation management, research policy, and the valuation of intellectual capital. He has carried out research and consultancy projects funded by companies, ministries, and the European Commission as well as taught innovation management at the Vienna University of Technology. His research has been published in *R&D Management, Creativity and Innovation Management,* and *the International Journal of Technology Management,* among others.

Martin Lindell is a professor in Entrepreneurship and Management at Hanken School of Economics. His main research interests are in innovations, leader behaviors, strategy, acquisitions, and corporate governance. He has published his work in books and journals, such as *Leadership Quarterly, Scandinavian Management Journal, International Studies of Management and Organization, Journal of Small Business Management,* and *Business Strategy and the Environment.*

Mitsutaka Matsumoto is a research scientist at the Center for Service Research at the National Institute of Advanced Industrial Science and Technology (AIST), Japan. His research focuses on innovation management, product life cycle engineering, and service engineering. He has published his work in such journals as *R&D Management, Journal of Cleaner Production,* and *Journal of Remanufacturing.*

M. Paloma Sánchez is a Professor of Applied Economics at the Autonomous University of Madrid, a Director of an Interuniversity Master and PhD Programme on Economics and Management of Innovation, and a holder of the UAM-Accenture Chair in Economics and Management of Innovation. She is chairperson of the Committee for Scientific and Technological Policy of the OECD (1990–1993) and member of several high level expert groups at the European Commission on innovation-related issues.

Beata Segercrantz is a researcher at Hanken School of Economics in Helsinki. She holds a PhD in Management and Organization (Hanken School of Economics) and a master's degree in Social Psychology (University of Helsinki). Her main research interests and topics of publications are in new product development, innovation, organizational change and restructurings, identity, subjectivity, gender, and discourse analysis.

Nikodemus Solitander is a university lecturer in Supply Chain Management and Corporate Geography at Hanken School of Economics, Helsinki. His main research interests are related to sustainable development, ethics, and creativity—sometimes relatedly, sometimes irrespectively. He has previously published in journals such as *Management Decision* and *Corporate Social Responsibility and Environmental Management* and has contributed to several edited books.

Karl-Erik Sveiby is a professor in Knowledge Management at Hanken School of Economics, Helsinki. His book *The New Organizational Wealth* (1997) is regarded as one of the seminal works in the field. He is member of several editorial boards; among others, *Journal of Intellectual Capital* and *Journal of Knowledge Management*. His latest works are inspired by indigenous knowledge: *Aboriginal Principles for a Sustainable Society—As Told in Traditional Law Stories* (*Sustainable Development*, 2009) and *Collective Leadership with Power Symmetry—Lessons from Aboriginal Prehistory* (*Journal of Leadership*, forthcoming). His current research interests are innovation and nonhierarchical leadership in jazz bands.

Urmas Varblane is a professor of International Business in the University of Tartu, Estonia. His research is focused on the knowledge transfer from FDI, industry-academia partnerships, and innovation systems. He has worked in the Augsburg University, Kiel Institute of World Economy, and University College of London. He has been involved in many research projects, for example, *Knowledge-Based Entrepreneurship: Innovation, Networks and Systems* and *Knowledge and Competitiveness in the Enlarged EU*. He is a member of editorial board at *Journal of Knowledge Economy, Industry and Higher Education, Economic Systems*, and *Journal of East West Business*.

Author Index

Subject Index